THE CONTEST PROBLEM BOOK II

Annual High School Contests
1961–1965
of
The Mathematical Association of America
Society of Actuaries
Mu Alpha Theta
National Council of Teachers of Mathematics
Casualty Actuarial Society

NEW MATHEMATICAL LIBRARY
PUBLISHED BY

THE MATHEMATICAL ASSOCIATION OF AMERICA

Editorial Committee

Ivan Niven, Chairman (1978-80)　　Anneli Lax, Editor
　University of Oregon　　　　　　*New York University*

W. G. Chinn (1977-79)　*City College of San Francisco*
Basil Gordon (1977-79)　*University of California, Los Angeles*
M. M. Schiffer (1976-78)　*Stanford University*

The New Mathematical Library (NML) was begun in 1961 by the School Mathematics Study Group to make available to high school students short expository books on various topics not usually covered in the high school syllabus. In a decade the NML matured into a steadily growing series of some twenty titles of interest not only to the originally intended audience, but to college students and teachers at all levels. Previously published by Random House and L. W. Singer, the NML became a publication series of the Mathematical Association of America (MAA) in 1975. Under the auspices of the MAA the NML will continue to grow and will remain dedicated to its original and expanded purposes.

THE CONTEST PROBLEM BOOK II

Annual High School Contests

1961–1965

compiled and with solutions by

Charles T. Salkind
Polytechnic Institute of Brooklyn

17

THE MATHEMATICAL ASSOCIATION
OF AMERICA

Illustrations by George H. Buehler

Ninth Printing

© Copyright, 1966 by The Mathematical Association of America (Inc.)

All Rights reserved under International and Pan-American Copyright Conventions. Published in Washington, D.C. by
The Mathematical Association of America

Library of Congress Catalog Card Number: 66-15479
Complete Set: ISBN-0-88385-600-X
Vol. 17: ISBN-0-88385-617-4

Manufactured in the United States of America

Contents

	Preface	3
	Editors' Preface	5
	Suggestions for Using this Book	7
I	Problems	9
	1961 Examination	9
	1962 Examination	16
	1963 Examination	23
	1964 Examination	31
	1965 Examination	38
II	Answer Keys	47
III	Solutions	49
	1961 Examination	49
	1962 Examination	58
	1963 Examination	69
	1964 Examination	85
	1965 Examination	96
IV	Classification of Problems	109

NEW MATHEMATICAL LIBRARY

1. Numbers: Rational and Irrational *by Ivan Niven*
2. What is Calculus About? *by W. W. Sawyer*
3. An Introduction to Inequalities *by E. F. Beckenbach and R. Bellman*
4. Geometric Inequalities *by N. D. Kazarinoff*
5. The Contest Problem Book I Annual High School Mathematics Examinations 1950–1960. Compiled and with solutions *by Charles T. Salkind*
6. The Lore of Large Numbers, *by P. J. Davis*
7. Uses of Infinity *by Leo Zippin*
8. Geometric Transformations I *by I. M. Yaglom, translated by A. Shields*
9. Continued Fractions *by Carl D. Olds*
10. Graphs and Their Uses *by Oystein Ore*
11. ⎫ Hungarian Problem Books I and II, Based on the Eötvös
12. ⎭ Competitions 1894–1905 and 1906–1928, *translated by E. Rapaport*
13. Episodes from the Early History of Mathematics *by A. Aaboe*
14. Groups and Their Graphs *by I. Grossman and W. Magnus*
15. The Mathematics of Choice or How to Count Without Counting *by Ivan Niven*
16. From Pythagoras to Einstein *by K. O. Friedrichs*
17. The Contest Problem Book II Annual High School Mathematics Examinations 1961–1965. Compiled and with solutions *by Charles T. Salkind*
18. First Concepts of Topology *by W. G. Chinn and N. E. Steenrod*
19. Geometry Revisited *by H. S. M. Coxeter and S. L. Greitzer*
20. Invitation to Number Theory *by Oystein Ore*
21. Geometric Transformations II *by I. M. Yaglom, translated by A. Shields*
22. Elementary Cryptanalysis—A Mathematical Approach *by A. Sinkov*
23. Ingenuity in Mathematics *by Ross Honsberger*
24. Geometric Transformations III *by I. M. Yaglom, translated by A. Shenitzer*
25. The Contest Problem Book III Annual High School Mathematics Examinations 1966–1972. Compiled and with solutions *by C. T. Salkind and J. M. Earl*
26. Mathematical Methods in Science *by George Pólya*
27. International Mathematical Olympiads—1959–1977. Compiled and with solutions *by S. L. Greitzer*
28. The Mathematics of Games and Gambling *by Edward W. Packel*
29. The Contest Problem Book IV Annual High School Mathematics Examinations 1973–1982. Compiled and with solutions *by R. A. Artino, A. M. Gaglione and N. Shell*
30. The Role of Mathematics in Science *by M. M. Schiffer and L. Bowden*
31. International Mathematical Olympiads 1978–1985 and forty supplementary problems. Compiled and with solutions *by Murray S. Klamkin*
32. Riddles of the Sphinx (and Other Mathematical Puzzle Tales) *by Martin Gardner*

Other titles in preparation.

THE CONTEST PROBLEM BOOK II

Annual High School Contests
1961–1965
of
The Mathematical Association of America
Society of Actuaries
Mu Alpha Theta
National Council of Teachers of Mathematics
Casualty Actuarial Society

Preface

The thesis that selective problem solving can be a vital factor in learning mathematics needs no extended defense. It is implicit in the suggestion by some curriculum experts that problems be made the central point of topical development. A good problem, like the acorn, has in it the potential for grand development.

The Committee on High School Contests, guided by this thesis, seeks through the Annual High School Mathematics Examination—jointly sponsored by the Mathematical Association of America and the Society of Actuaries—to extend, to supplement, and to enrich the regular school work by providing interesting, rewarding, and challenging problems within a prescribed scope.

There is one very important respect in which this competition differs from the Olympiads of Europe and similarly-motivated competitions in the United States and Canada. This competition aims to discriminate on *several levels*, and is not *exclusively* directed to high-ability students.

Since the publication of the contest problems through 1960 (NML volume 5), participation in this contest has increased by 100,000 students in the United States and Canada to its present 250,000 from 7000 schools. There is also a fairly large European participation.

Some of the solutions are intentionally incomplete but crucial steps are shown. The exhibited solutions are by no means the only ones possible, nor are they necessarily superior to all alternatives. Since it is our intention that no mathematics beyond intermediate algebra be required we consistently show an elementary procedure even where a "high-powered" alternative is given.

Your comments are invited.

Charles T. Salkind

Editors' Preface

The editors of the New Mathematical Library, in wishing to encourage significant problem-solving at the high school level, have published the following problem collections so far: NML 5 contains all problems proposed through 1960 by the Mathematical Association of America in its annual contests for high school students. NML 11 and NML 12 contain translations of all Eötvös Competition problems through 1928 and their solutions. The present volume is a sequel to NML 5, published at the request of the many readers who enjoyed that book. Sequels to the Hungarian Problem Books are also contemplated by·the NML.

The MAA contest is based entirely on the standard high school curriculum and contains forty problems. (Before 1960, there were fifty problems in each contest.) Each Eötvös contest, on the other hand, contains only three problems, also based on the (Hungarian) high school curriculum, but requiring ingenuity and often rather deep investigation for their solution.

The MAA is concerned primarily with mathematics on the undergraduate level. It is one of the three major mathematical organizations in America (the other two being the American Mathematical Society, chiefly concerned with mathematical research, and the National Council of Teachers of Mathematics, concerned with the content and pedagogy of elementary and secondary mathematics). The MAA also conducts the annual "Putnam Competition" for undergraduate students. Its journal, *The American Mathematical Monthly*, is famous for its elementary and advanced problem sections.

When the MAA contest was first organized in 1950 it was restricted to the metropolitan New York area. It became a national project in 1957, receiving in that same year co-sponsorship by the Society of Actuaries. In 1960 more than 150,000 students from 5,200 schools participated in the program. The contest is conducted in nearly every state and territory of the U. S. and in the more populated provinces of Canada.

Part I of each examination tests fundamental skills based on conceptual understanding, while Parts II and III probe beyond mere reproduction of classroom work. For the ten years this contest was conducted, only three people scored† perfectly (150); a score of 80 or more places a contestant on the Honor Roll.

The editors are grateful to the MAA for permission to publish this collection, and to Professor Charles T. Salkind for compiling the book and for supplying a classification of problems with their complete solutions. In preparing this collection, the compiler and the editors have made a few minor changes in the statements of the original contest problems for the sake of greater clarity.

L. Bers
J. H. Hlavaty
New York, 1965

---

† Scoring is as follows: 1950–1959, Part I, 2 points each; Part II, 3 points each; Part III, 4 points each. 1960–1965, Part I, 3 points each; Part II, 4 points each; Part III, 5 points each.

Suggestions for Using this Book

This problem collection is designed to be used by mathematics clubs, high school teachers, students, and other interested individuals. Clearly, no one would profit from doing *all* the problems, but he *would* benefit from those that present a challenge to him. The reader might try himself on a whole test or on part of a test, with (or preferably without) time limitations.

He should try to get as far as possible with the solution to a problem. If he is really stuck, he should look up the answer in the key, see page 47, and try to work backwards; if this fails, the section of complete solutions should be consulted, see pages 49–107.

In studying solutions, even the successful problem solver may find sidelights he had overlooked; he may find a more "elegant" solution, or a way of solving the problem which may lead him deeper into mathematics. He may find it interesting to change items in the hypothesis and to see how this affects the solution, or to invent his own problems.

If a reader is interested in a special type of problem, he should consult the classified index.

The following familiar symbols appear in this book:

Symbol	Meaning
$\sim$	similar (if used in connection with plane figures)
$\sim$	approximately equal (if used in connection with numbers)
$\therefore$	therefore
$\equiv$	identically equal to
$<$	less than
$\leq$	less than or equal to
$>$	greater than
$\geq$	greater than or equal to
$\lvert k \rvert$	absolute value of the number k
$\triangle$	triangle
110_2	the number $1 \cdot 2^2 + 1 \cdot 2^1 + 0 \cdot 2^0$, i.e., the number 6 when written in a numeration system with base 2 instead of 10.
$\cong$	congruent
$\neq$	different from
$\perp$	perpendicular to
XY	length of the line segment XY, often denoted by $\overline{XY}$ in other books
$f(x)$	function of the variable x

I
Problems

1961 Examination

Part 1

1. When simplified $(-\frac{1}{125})^{-2/3}$ becomes:

   (A) $\frac{1}{25}$ (B) $-\frac{1}{25}$ (C) 25 (D) -25 (E) $25\sqrt{-1}$

2. An automobile travels $a/6$ *feet* in r *seconds*. If this rate is maintained for 3 minutes, how many *yards* does it travel in the 3 *minutes*?

   (A) $\dfrac{a}{1080r}$ (B) $\dfrac{30r}{a}$ (C) $\dfrac{30a}{r}$ (D) $\dfrac{10r}{a}$ (E) $\dfrac{10a}{r}$

3. If the graphs of $2y + x + 3 = 0$ and $3y + ax + 2 = 0$ are to meet at right angles, the value of a is:

   (A) $\pm\frac{2}{3}$ (B) $-\frac{2}{3}$ (C) $-\frac{3}{2}$ (D) 6 (E) -6

4. Let the set consisting of the squares of the positive integers be called u; thus u is the set $1, 4, 9, \cdots$. If a certain operation on one or more members of the set always yields a member of the set, we say that the set is closed under that operation. Then u is closed under:

   (A) addition   (B) multiplication   (C) division
   (D) extraction of a positive integral root   (E) none of these

5. Let $S = (x-1)^4 + 4(x-1)^3 + 6(x-1)^2 + 4(x-1) + 1$. Then S equals:

   (A) $(x-2)^4$ (B) $(x-1)^4$ (C) x^4 (D) $(x+1)^4$ (E) x^4+1

6. When simplified, $\log 8 \div \log \tfrac{1}{8}$ becomes:

   (A) $6 \log 2$ (B) $\log 2$ (C) 1 (D) 0 (E) -1

7. When simplified, the third term in the expansion of $\left(\dfrac{a}{\sqrt{x}} - \dfrac{\sqrt{x}}{a^2} \right)^6$ is:

   (A) $\dfrac{15}{x}$ (B) $-\dfrac{15}{x}$ (C) $-\dfrac{6x^2}{a^9}$ (D) $\dfrac{20}{a^3}$ (E) $-\dfrac{20}{a^3}$

8. Let the two base angles of a triangle be A and B, with B larger than A. The altitude to the base divides the vertex angle C into two parts, C_1 and C_2, with C_2 adjacent to side a. Then:

   (A) $C_1 + C_2 = A + B$ (B) $C_1 - C_2 = B - A$
   (C) $C_1 - C_2 = A - B$ (D) $C_1 + C_2 = B - A$
   (E) $C_1 - C_2 = A + B$

9. Let r be the result of doubling both the base and the exponent of a^b, $b \neq 0$. If r equals the product of a^b by x^b, then x equals:

   (A) a (B) $2a$ (C) $4a$ (D) 2 (E) 4

10. Each side of triangle ABC is 12 units. D is the foot of the perpendicular dropped from A on BC, and E is the midpoint of AD. The length of BE, in the same unit, is:

    (A) $\sqrt{18}$ (B) $\sqrt{28}$ (C) 6 (D) $\sqrt{63}$ (E) $\sqrt{98}$

11. Two tangents are drawn to a circle from an exterior point A; they touch the circle at points B and C, respectively. A third tangent intersects segment AB in P and AC in R, and touches the circle at Q. If $AB = 20$, then the perimeter of triangle APR is:

    (A) 42 (B) 40.5 (C) 40 (D) $39\tfrac{7}{8}$
    (E) not determined by the given information

12. The first three terms of a geometric progression are $\sqrt{2}, \sqrt[3]{2}, \sqrt[6]{2}$. The fourth term is:

   (A) 1   (B) $\sqrt[7]{2}$ (C) $\sqrt[8]{2}$ (D) $\sqrt[9]{2}$ (E) $\sqrt[10]{2}$

13. The symbol $|a|$ means a if a is a positive number or zero, and $-a$ if a is a negative number. For all real values of t the expression $\sqrt{t^4 + t^2}$ is equal to:

   (A) t^3 (B) $t^2 + t$ (C) $|t^2 + t|$ (D) $t\sqrt{t^2+1}$
   (E) $|t|\sqrt{1+t^2}$

14. A rhombus is given with one diagonal twice the length of the other diagonal. Express the side of the rhombus in terms of K, where K is the area of the rhombus in square inches.

   (A) $\sqrt{K}$ (B) $\tfrac{1}{2}\sqrt{2K}$ (C) $\tfrac{1}{3}\sqrt{3K}$ (D) $\tfrac{1}{4}\sqrt{4K}$
   (E) none of these is correct

15. If x men working x hours a day for each of x days produce x articles, then the number of articles (not necessarily an integer) produced by y men working y hours a day for each of y days is:

   (A) $\dfrac{x^3}{y^2}$ (B) $\dfrac{y^3}{x^2}$ (C) $\dfrac{x^2}{y^3}$ (D) $\dfrac{y^2}{x^3}$ (E) y

16. An altitude h of a triangle is increased by a length m. How much must be taken from the corresponding base b so that the area of the new triangle is one-half that of the original triangle?

   (A) $\dfrac{bm}{h+m}$ (B) $\dfrac{bh}{2(h+m)}$ (C) $\dfrac{b(2m+h)}{m+h}$
   (D) $\dfrac{b(m+h)}{2m+h}$ (E) $\dfrac{b(2m+h)}{2(h+m)}$

17. In the base ten number system the number 526 means $5 \cdot 10^2 + 2 \cdot 10 + 6$. In the Land of Mathesis, however, numbers are written in the base r. Jones purchases an automobile there for 440 monetary units (abbreviated m.u.). He gives the salesman a 1000 m.u. bill, and receives, in change, 340 m.u. The base r is:

   (A) 2   (B) 5   (C) 7   (D) 8   (E) 12

18. The yearly changes in the population census of a town for four consecutive years are, respectively, 25% increase, 25% increase, 25% decrease, 25% decrease. The net change over the four years, to the nearest percent, is:

   (A) -12 (B) -1 (C) 0   (D) 1   (E) 12

19. Consider the graphs of $y = 2 \log x$ and $y = \log 2x$. We may say that:

   (A) They do not intersect.
   (B) They intersect in one point only.
   (C) They intersect in two points only.
   (D) They intersect in a finite number of points but more than two.
   (E) They coincide.

20. The set of points satisfying the pair of inequalities $y > 2x$ and $y > 4 - x$ is contained entirely in quadrants:

   (A) I and II   (B) II and III   (C) I and III
   (D) III and IV   (E) I and IV

Part 2

21. Medians AD and CE of triangle ABC intersect in M. The midpoint of AE is N. Let the area of triangle MNE be k times the area of triangle ABC. Then k equals:

   (A) $\frac{1}{6}$ (B) $\frac{1}{8}$ (C) $\frac{1}{9}$ (D) $\frac{1}{12}$ (E) $\frac{1}{16}$

22. If $3x^3 - 9x^2 + kx - 12$ is divisible by $x - 3$, then it is also divisible by:

   (A) $3x^2 - x + 4$ (B) $3x^2 - 4$ (C) $3x^2 + 4$
   (D) $3x - 4$ (E) $3x + 4$

23. Points P and Q are both in the line segment AB and on the same side of its midpoint. P divides AB in the ratio $2:3$, and Q divides AB in the ratio $3:4$. If $PQ = 2$, then the length of AB is:

   (A) 60   (B) 70   (C) 75   (D) 80   (E) 85

PROBLEMS: 1961 EXAMINATION   13

24. Thirty-one books are arranged from left to right in order of increasing prices. The price of each book differs by $2 from that of each adjacent book. For the price of the book at the extreme right a customer can buy the middle book and an adjacent one. Then:

   (A) The adjacent book referred to is at the left of the middle book.
   (B) The middle book sells for $36.
   (C) The cheapest book sells for $4.
   (D) The most expensive book sells for $64.
   (E) None of these is correct.

25. Triangle ABC is isosceles with base AC. Points P and Q are respectively in CB and AB and such that $AC = AP = PQ = QB$. The number of degrees in angle B is:

   (A) $25\frac{5}{7}$ (B) $26\frac{1}{3}$ (C) 30   (D) 40
   (E) not determined by the information given

26. For a given arithmetic series the sum of the first 50 terms is 200, and the sum of the next 50 terms is 2700. The first term of the series is:

   (A) -1221 (B) -21.5 (C) -20.5 (D) 3   (E) 3.5

27. Given two equiangular polygons P_1 and P_2 with different numbers of sides; each angle of P_1 is x degrees and each angle of P_2 is kx degrees, where k is an integer greater than 1. The number of possibilities for the pair (x, k) is:

   (A) infinite   (B) finite, but more than two
   (C) two   (D) one   (E) zero

28. If 2137^{753} is multiplied out, the units' digit in the final product is:

   (A) 1   (B) 3   (C) 5   (D) 7   (E) 9

29. Let the roots of $ax^2 + bx + c = 0$ be r and s. The equation with roots $ar + b$ and $as + b$ is:

   (A) $x^2 - bx - ac = 0$ (B) $x^2 - bx + ac = 0$
   (C) $x^2 + 3bx + ca + 2b^2 = 0$ (D) $x^2 + 3bx - ca + 2b^2 = 0$
   (E) $x^2 + bx(2 - a) + a^2c + b^2(a + 1) = 0$

30. If $\log_{10} 2 = a$ and $\log_{10} 3 = b$, then $\log_5 12$ equals:

(A) $\dfrac{a+b}{1+a}$ (B) $\dfrac{2a+b}{1+a}$ (C) $\dfrac{a+2b}{1+a}$ (D) $\dfrac{2a+b}{1-a}$ (E) $\dfrac{a+2b}{1-a}$

Part 3

31. In triangle ABC the ratio $AC:CB$ is $3:4$. The bisector of the exterior angle at C intersects BA extended at P (A is between P and B). The ratio $PA:AB$ is:

(A) $1:3$ (B) $3:4$ (C) $4:3$ (D) $3:1$ (E) $7:1$

32. A regular polygon of n sides is inscribed in a circle of radius R. The area of the polygon is $3R^2$. Then n equals:

(A) 8 (B) 10 (C) 12 (D) 15 (E) 18

33. The number of solutions of $2^{2x} - 3^{2y} = 55$, in which x and y are integers, is:

(A) zero (B) one (C) two (D) three
(E) more than three, but finite

34. Let S be the set of values assumed by the function

$$\frac{2x+3}{x+2}$$

when x is any member of the interval $x \geq 0$. If there exists a number M such that no number of the set S is greater than M, then M is an upper bound of S. If there exists a number m such that no number of the set S is less than m, then m is a lower bound of S. We may then say:

(A) m is in S, M is not in S.
(B) M is in S, m is not in S.
(C) Both m and M are in S.
(D) Neither m nor M is in S.
(E) M does not exist either in or outside S.

35. The number 695 is to be written with a factorial base of numeration, that is, $695 = a_1 + a_2 \cdot 2! + a_3 \cdot 3! + \cdots + a_n \cdot n!$ where $a_1, a_2, \cdots, a_n$ are

integers such that $0 \leq a_k \leq k$, and $n!$ means $n(n-1)(n-2) \cdots 2 \cdot 1$. Find a_4.

(A) 0   (B) 1   (C) 2   (D) 3   (E) 4

36. In triangle ABC the median from A is given perpendicular to the median from B. If $BC = 7$ and $AC = 6$, then the length of AB is:

(A) 4   (B) $\sqrt{17}$ (C) 4.25   (D) $2\sqrt{5}$ (E) 4.5

37. In racing over a given distance d at uniform speed, A can beat B by 20 yards, B can beat C by 10 yards, and A can beat C by 28 yards. Then d, in yards, equals:

(A) not determined by the given information
(B) 58   (C) 100   (D) 116   (E) 120

38. Triangle ABC is inscribed in a semicircle of radius r so that its base AB coincides with diameter AB. Point C does not coincide with either A or B. Let $s = AC + BC$. Then, for all permissible positions of C:

(A) $s^2 \leq 8r^2$ (B) $s^2 = 8r^2$ (C) $s^2 \geq 8r^2$
(D) $s^2 \leq 4r^2$ (E) $s^2 = 4r^2$

39. Any five points are taken inside or on a square of side 1. Let a be the *smallest* possible number with the property that it is always possible to select one pair of points from these five such that the distance between them is equal to or less than a. Then a is:

(A) $\sqrt{3}/3$ (B) $\sqrt{2}/2$ (C) $2\sqrt{2}/3$ (D) 1   (E) $\sqrt{2}$

40. Find the minimum value of $\sqrt{x^2 + y^2}$ if $5x + 12y = 60$.

(A) $\frac{60}{13}$ (B) $\frac{13}{5}$ (C) $\frac{13}{12}$ (D) 1   (E) 0

1962 Examination

Part 1

1. The expression $\dfrac{1^{4y-1}}{5^{-1}+3^{-1}}$ is equal to:

   (A) $\dfrac{4y-1}{8}$ (B) 8 (C) $\dfrac{15}{2}$ (D) $\dfrac{15}{8}$ (E) $\dfrac{1}{8}$

2. The expression $\sqrt{\tfrac{4}{3}} - \sqrt{\tfrac{3}{4}}$ is equal to:

   (A) $\sqrt{3}/6$ (B) $-\sqrt{3}/6$ (C) $\sqrt{-3}/6$ (D) $5\sqrt{3}/6$ (E) 1

3. The first three terms of an arithmetic progression are $x-1$, $x+1$, $2x+3$, in the order shown. The value of x is:

   (A) -2 (B) 0 (C) 2 (D) 4 (E) undetermined

4. If $8^x = 32$, then x equals:

   (A) 4 (B) $\tfrac{5}{3}$ (C) $\tfrac{3}{2}$ (D) $\tfrac{3}{5}$ (E) $\tfrac{1}{4}$

5. If the radius of a circle is increased by 1 unit, the ratio of the new circumference to the new diameter is:

   (A) $\pi + 2$ (B) $\dfrac{2\pi+1}{2}$ (C) π (D) $\dfrac{2\pi-1}{2}$ (E) $\pi - 2$

6. A square and an equilateral triangle have equal perimeters. The area of the triangle is $9\sqrt{3}$ square inches. Expressed in inches the diagonal of the square is:

   (A) $9/2$ (B) $2\sqrt{5}$ (C) $4\sqrt{2}$ (D) $9\sqrt{2}/2$ (E) none of these

7. Let the bisectors of the exterior angles at B and C of triangle ABC meet at D. Then the measure in degrees of angle BDC is:

   (A) $\tfrac{1}{2}(90 - A)$ (B) $90 - A$ (C) $\tfrac{1}{2}(180 - A)$
   (D) $180 - A$ (E) $180 - 2A$

8. Given the set of n numbers, $n > 1$, of which one is $1 - (1/n)$, and all the others are 1. The arithmetic mean of the n numbers is:

(A) 1 (B) $n - \dfrac{1}{n}$ (C) $n - \dfrac{1}{n^2}$ (D) $1 - \dfrac{1}{n^2}$ (E) $1 - \dfrac{1}{n} - \dfrac{1}{n^2}$

9. When $x^9 - x$ is factored as completely as possible into polynomials and monomials with real integral coefficients, the number of factors is:

(A) more than 5 (B) 5 (C) 4 (D) 3 (E) 2

10. A man drives 150 miles to the seashore in 3 hours and 20 minutes. He returns from the shore to the starting point in 4 hours and 10 minutes. Let r be the average rate for the entire trip. Then the average rate for the trip going exceeds r, in miles per hour, by:

(A) 5 (B) $4\tfrac{1}{2}$ (C) 4 (D) 2 (E) 1

11. The difference between the larger root and the smaller root of
$$x^2 - px + (p^2 - 1)/4 = 0$$
is

(A) 0 (B) 1 (C) 2 (D) p (E) $p + 1$

12. When $\left(1 - \dfrac{1}{a}\right)^6$ is expanded, the sum of the last three coefficients is:

(A) 22 (B) 11 (C) 10 (D) -10 (E) -11

13. R varies directly as S and inversely as T. When $R = \tfrac{4}{3}$ and $T = \tfrac{9}{14}$, $S = \tfrac{3}{7}$. Find S when $R = \sqrt{48}$ and $T = \sqrt{75}$.

(A) 28 (B) 30 (C) 40 (D) 42 (E) 60

14. Let s be the limiting sum of the geometric series $4 - \tfrac{8}{3} + \tfrac{16}{9} - \cdots$, as the number of terms increases without bound. Then s equals:

(A) a number between 0 and 1
(B) 2.4 (C) 2.5 (D) 3.6 (E) 12

15. Given triangle ABC with base AB fixed in length and position. As the vertex C moves on a straight line, the intersection point of the three medians moves on:

(A) a circle (B) a parabola (C) an ellipse
(D) a straight line (E) a curve not listed here

16. Given rectangle R_1 with one side 2 inches and area 12 square inches. Rectangle R_2 with diagonal 15 inches is similar to R_1. Expressed in square inches the area of R_2 is:

   (A) 9/2 (B) 36 (C) 135/2 (D) $9\sqrt{10}$ (E) $27\sqrt{10}/4$

17. If $a = \log_8 225$ and $b = \log_2 15$, then a, in terms of b, is:

   (A) $b/2$ (B) $2b/3$ (C) b (D) $3b/2$ (E) $2b$

18. A regular dodecagon (12 sides) is inscribed in a circle with radius r inches. The area of the dodecagon, in square inches, is:

   (A) $3r^2$ (B) $2r^2$ (C) $3r^2\sqrt{3}/4$ (D) $r^2\sqrt{3}$ (E) $3r^2\sqrt{3}$

19. If the parabola $y = ax^2 + bx + c$ passes through the points $(-1, 12)$, $(0, 5)$, and $(2, -3)$, the value of $a + b + c$ is:

   (A) -4 (B) -2 (C) 0 (D) 1 (E) 2

20. The angles of a pentagon are in arithmetic progression. One of the angles, in degrees, must be:

   (A) 108 (B) 90 (C) 72 (D) 54 (E) 36

Part 2

21. It is given that one root of $2x^2 + rx + s = 0$, with r and s real numbers, is $3 + 2i$ ($i = \sqrt{-1}$). The value of s is:

   (A) undetermined (B) 5 (C) 6 (D) -13 (E) 26

22. The number 121_b, written in the integral base b, is the square of an integer for

   (A) $b = 10$, only (B) $b = 10$ and $b = 5$, only (C) $2 \leq b \leq 10$
   (D) $b > 2$ (E) no value of b

23. In triangle ABC, CD is the altitude to AB, and AE is the altitude to BC. If the lengths of AB, CD, and AE are known, the length of DB is:

(A) not determined by the information given
(B) determined only if A is an acute angle
(C) determined only if B is an acute angle
(D) determined only if ABC is an acute triangle
(E) none of these is correct

24. Three machines P, Q, and R, working together, can do a job in x hours. When working alone P needs an additional 6 hours to do the job; Q, one additional hour; and R, x additional hours. The value of x is:

    (A) $\frac{2}{3}$ (B) $\frac{11}{12}$ (C) $\frac{3}{2}$ (D) 2   (E) 3

25. Given square $ABCD$ with side 8 feet. A circle is drawn through vertices A and D and tangent to side BC. The radius of the circle, in feet, is:

    (A) 4   (B) $4\sqrt{2}$ (C) 5   (D) $5\sqrt{2}$ (E) 6

26. For any real value of x the maximum value of $8x - 3x^2$ is:

    (A) 0   (B) $\frac{8}{3}$ (C) 4   (D) 5   (E) $\frac{16}{3}$

27. Let $a \, \text{\textcircled{L}} \, b$ represent the operation on two numbers, a and b, which selects the larger of the two numbers, with $a \, \text{\textcircled{L}} \, a = a$. Let $a \, \text{\textcircled{S}} \, b$ represent the operation which selects the smaller of the two numbers, with $a \, \text{\textcircled{S}} \, a = a$.
    Which of the following three rules is (are) correct?

    (1) $a \, \text{\textcircled{L}} \, b = b \, \text{\textcircled{L}} \, a$, (2) $a \, \text{\textcircled{L}} \, (b \, \text{\textcircled{L}} \, c) = (a \, \text{\textcircled{L}} \, b) \, \text{\textcircled{L}} \, c$,
    (3) $a \, \text{\textcircled{S}} \, (b \, \text{\textcircled{L}} \, c) = (a \, \text{\textcircled{S}} \, b) \, \text{\textcircled{L}} \, (a \, \text{\textcircled{S}} \, c)$.

    (A) (1) only   (B) (2) only   (C) (1) and (2) only
    (D) (1) and (3) only   (E) all three

28. The set of x-values satisfying the equation $x^{\log_{10} x} = x^3/100$ consists of:

    (A) $\frac{1}{10}$, only   (B) 10, only   (C) 100, only   (D) 10 or 100, only
    (E) more than two real numbers.

29. Which of the following sets of x-values satisfy the inequality
$$2x^2 + x < 6 \, ?$$

    (A) $-2 < x < \frac{3}{2}$ (B) $x > \frac{3}{2}$ or $x < -2$
    (C) $x < \frac{3}{2}$ (D) $\frac{3}{2} < x < 2$ (E) $x < -2$

30. **Form I**

Consider the statements: (1) $p \wedge q$ (2) $p \wedge \sim q$ (3) $\sim p \wedge q$ (4) $\sim p \wedge \sim q$, where p and q are statements.
    How many of these imply the truth of $\sim(p \wedge q)$?

(A) 0   (B) 1   (C) 2   (D) 3   (E) 4

Form II

Consider the statements: (1) p and q are both true (2) p is true and q is false (3) p is false and q is true (4) p is false and q is false.
    How many of these imply the negation of the statement "p and q are both true"?

(A) 0   (B) 1   (C) 2   (D) 3   (E) 4

Part 3

31. The ratio of the interior angles of two regular polygons is 3:2. How many such pairs are there?

(A) 1   (B) 2   (C) 3   (D) 4   (E) infinitely many

32. If $x_{k+1} = x_k + \frac{1}{2}$ for $k = 1, 2, \cdots, n-1$ and $x_1 = 1$, find
$$x_1 + x_2 + \cdots + x_n.$$

(A) $\dfrac{n+1}{2}$ (B) $\dfrac{n+3}{2}$ (C) $\dfrac{n^2-1}{2}$ (D) $\dfrac{n^2+n}{4}$ (E) $\dfrac{n^2+3n}{4}$

33. The set of x-values satisfying the inequality $2 \leq |x-1| \leq 5$ is:

(A) $-4 \leq x \leq -1$ or $3 \leq x \leq 6$
(B) $3 \leq x \leq 6$ or $-6 \leq x \leq -3$
(C) $x \leq -1$ or $x \geq 3$
(D) $-1 \leq x \leq 3$
(E) $-4 \leq x \leq 6$

34. For what real values of K does $x = K^2(x-1)(x-2)$ have real roots?

(A) none   (B) $-2 < K < 1$ (C) $-2\sqrt{2} < K < 2\sqrt{2}$
(D) $K > 1$ or $K < -2$ (E) all

35. A man on his way to dinner shortly after 6:00 p.m. observes that the hands of his watch form an angle of 110°. Returning before 7:00 p.m. he notices that again the hands of his watch form an angle of 110°. The number of minutes that he has been away is:

(A) $36\tfrac{2}{3}$ (B) 40   (C) 42   (D) 42.4   (E) 45

36. If both x and y are integers, how many solutions are there of the equation $(x-8)(x-10) = 2^y$?

(A) 0   (B) 1   (C) 2   (D) 3   (E) more than 3

37. $ABCD$ is a square with side of unit length. Points E and F are taken respectively on sides AB and AD so that $AE = AF$ and the quadrilateral $CDFE$ has maximum area. In square units this maximum area is:

(A) $\tfrac{1}{2}$ (B) $\tfrac{9}{16}$ (C) $\tfrac{19}{32}$ (D) $\tfrac{5}{8}$ (E) $\tfrac{2}{3}$

38. The population of Nosuch Junction at one time was a perfect square. Later, with an increase of 100, the population was one more than a perfect square. Now, with an additional increase of 100, the population is again a perfect square. The original population is a multiple of:

(A) 3   (B) 7   (C) 9   (D) 11   (E) 17

39. The medians AN and BP of a triangle with unequal sides are, respectively, 3 inches and 6 inches long. Its area is $3\sqrt{15}$ square inches. The length of the third median, in inches, is:

(A) 4   (B) $3\sqrt{3}$ (C) $3\sqrt{6}$ (D) $6\sqrt{3}$ (E) $6\sqrt{6}$

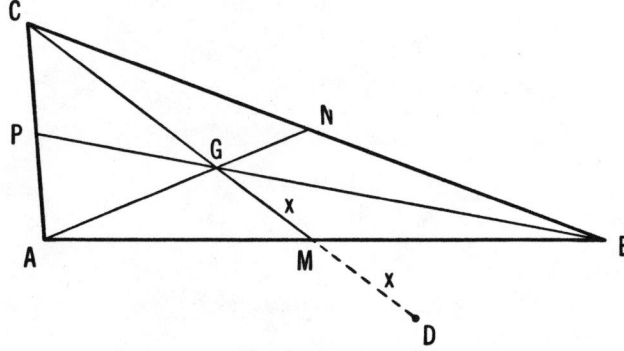

This diagram contains a hint for the solution.

40. The limiting sum of the infinite series
$$\frac{1}{10}+\frac{2}{10^2}+\frac{3}{10^3}+\cdots$$
whose n-th term is $n/10^n$ is:

(A) $\frac{1}{9}$ (B) $\frac{10}{81}$ (C) $\frac{1}{8}$ (D) $\frac{17}{72}$
(E) larger than any finite quantity

1963 Examination

Part 1

1. Which one of the following points is *not* on the graph of $y = \dfrac{x}{x+1}$?

   (A) $(0, 0)$ (B) $(-\tfrac{1}{2}, -1)$ (C) $(\tfrac{1}{2}, \tfrac{1}{3})$
   (D) $(-1, 1)$ (E) $(-2, 2)$

2. Let $n = x - y^{x-y}$. Find n when $x = 2$ and $y = -2$.

   (A) -14 (B) 0 (C) 1 (D) 18 (E) 256

3. If the reciprocal of $x + 1$ is $x - 1$, then x equals:

   (A) 0 (B) 1 (C) -1 (D) $+1$ or -1 (E) none of these

4. For what value(s) of k does the pair of equations $y = x^2$ and $y = 3x + k$ have two identical solutions?

   (A) $\tfrac{4}{9}$ (B) $-\tfrac{4}{9}$ (C) $\tfrac{9}{4}$ (D) $-\tfrac{9}{4}$ (E) $\tfrac{9}{4}$ or $-\tfrac{9}{4}$

5. If x and $\log_{10} x$ are real numbers and $\log_{10} x < 0$, then:

   (A) $x < 0$ (B) $-1 < x < 1$ (C) $0 < x \leq 1$
   (D) $-1 < x < 0$ (E) $0 < x < 1$

6. Triangle ABD is right-angled at B. On AD there is a point C for which $AC = CD$ and $AB = BC$. The magnitude of angle DAB, in degrees, is:

   (A) $67\tfrac{1}{2}$ (B) 60 (C) 45 (D) 30 (E) $22\tfrac{1}{2}$

7. Given the four equations:

   (1) $3y - 2x = 12$,   (2) $-2x - 3y = 10$,
   (3) $3y + 2x = 12$,   (4) $2y + 3x = 10$.

   The pair representing perpendicular lines is:

   (A) (1) and (4)   (B) (1) and (3)   (C) (1) and (2)
   (D) (2) and (4)   (E) (2) and (3)

8. The smallest positive integer x for which $1260x = N^3$, where N is an integer, is:

(A) 1050   (B) 1260   (C) 1260^2 (D) 7350   (E) 44,100

9. In the expansion of $\left(a - \dfrac{1}{\sqrt{a}}\right)^7$ the coefficient of $a^{-1/2}$ is:

(A) -7 (B) 7   (C) -21 (D) 21   (E) 35

10. Point P is taken interior to a square of side-length a and such that it is equally distant from two consecutive vertices and from the side opposite these vertices. If d represents the common distance, then d equals:

(A) $\dfrac{3a}{5}$ (B) $\dfrac{5a}{8}$ (C) $\dfrac{3a}{8}$ (D) $\dfrac{a\sqrt{2}}{2}$ (E) $\dfrac{a}{2}$

11. The arithmetic mean of a set of 50 numbers is 38. If two numbers of the set, namely 45 and 55, are discarded, the arithmetic mean of the remaining set of numbers is:

(A) 38.5   (B) 37.5   (C) 37   (D) 36.5   (E) 36

12. Three vertices of parallelogram $PQRS$ are $P(-3, -2)$, $Q(1, -5)$, $R(9, 1)$ with P and R diagonally opposite. The sum of the coordinates of vertex S is:

(A) 13   (B) 12   (C) 11   (D) 10   (E) 9

13. If $2^a + 2^b = 3^c + 3^d$, how many of the integers a, b, c, d can be negative?

(A) 4   (B) 3   (C) 2   (D) 1   (E) 0

14. Given the equations $x^2 + kx + 6 = 0$ and $x^2 - kx + 6 = 0$. If, when the roots of the equations are suitably listed, each root of the second equation is 5 more than the corresponding root of the first equation, then k equals:

(A) 5   (B) -5 (C) 7   (D) -7 (E) none of these

15. A circle is inscribed in an equilateral triangle, and a square is inscribed in the circle. The ratio of the area of the triangle to the area of the square is:

(A) $\sqrt{3}:1$ (B) $\sqrt{3}:\sqrt{2}$ (C) $3\sqrt{3}:2$ (D) $3:\sqrt{2}$ (E) $3:2\sqrt{2}$

16. Three numbers a, b, c, none zero, form an arithmetic progression. Increasing a by 1 or increasing c by 2 results in a geometric progression. Then b equals:

(A) 16   (B) 14   (C) 12   (D) 10   (E) 8

17. The expression

$$\left(\frac{a}{a+y}+\frac{y}{a-y}\right)\bigg/\left(\frac{y}{a+y}-\frac{a}{a-y}\right)$$

a real, $a \neq 0$, has the value -1 for:

(A) all but two real values of y (B) only two real values of y
(C) all real values of y (D) only one real value of y
(E) no real values of y

18. Chord EF is the perpendicular bisector of chord BC, intersecting it in M. Between B and M point U is taken, and EU extended meets the circle in A. Then, for any selection of U, as described, triangle EUM is similar to triangle:

(A) EFA (B) EFC (C) ABM (D) ABU (E) FMC

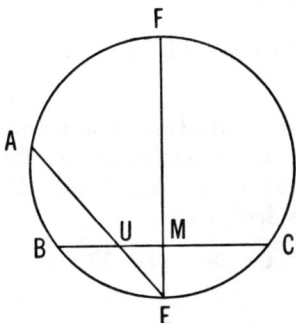

19. In counting n colored balls, some red and some black, it was found that 49 of the first 50 counted were red. Thereafter, 7 out of every 8 counted were red. If, in all, 90% or more of the balls counted were red, the maximum value of n is:

(A) 225   (B) 210   (C) 200   (D) 180   (E) 175

20. Two men at points R and S, 76 miles apart, set out at the same time to walk towards each other. The man at R walks uniformly at the rate of $4\frac{1}{2}$ miles per hour; the man at S walks at the constant rate of $3\frac{1}{4}$ miles per hour for the first hour, at $3\frac{3}{4}$ miles per hour for the second hour, and so on, in arithmetic progression. If the men meet x miles nearer R than S in an integral number of hours, then x is:

(A) 10   (B) 8   (C) 6   (D) 4   (E) 2

Part 2

21. The expression $x^2 - y^2 - z^2 + 2yz + x + y - z$ has:

(A) no linear factor with integer coefficients and integer exponents
(B) the factor $-x + y + z$
(C) the factor $x - y - z + 1$
(D) the factor $x + y - z + 1$
(E) the factor $x - y + z + 1$

22. Acute-angled triangle ABC is inscribed in a circle with center at O; $\widehat{AB} = 120°$ and $\widehat{BC} = 72°$. A point E is taken in minor arc AC such that OE is perpendicular to AC. Then the ratio of the magnitudes of angles OBE and BAC is:

(A) $\frac{5}{18}$ (B) $\frac{2}{9}$ (C) $\frac{1}{4}$ (D) $\frac{1}{3}$ (E) $\frac{4}{9}$

23. A gives B as many cents as B has and C as many cents as C has. Similarly, B then gives A and C as many cents as each then has. C, similarly, then gives A and B as many cents as each then has. If each finally has 16 cents, with how many cents does A start?

(A) 24   (B) 26   (C) 28   (D) 30   (E) 32

24. Consider equations of the form $x^2 + bx + c = 0$. How many such equations have real roots and have coefficients b and c selected from the set of integers $\{1, 2, 3, 4, 5, 6\}$?

(A) 20   (B) 19   (C) 18   (D) 17   (E) 16

25. Point F is taken in side AD of square $ABCD$. At C a perpendicular is drawn to CF, meeting AB extended at E. The area of $ABCD$ is 256

square inches and the area of triangle CEF is 200 square inches. Then the number of inches in BE is:

(A) 12   (B) 14   (C) 15   (D) 16   (E) 20

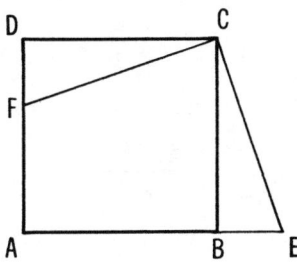

26. **Form I**

    Consider the statements

    (1) $p \wedge \sim q \wedge r$ (2) $\sim p \wedge \sim q \wedge r$
    (3) $p \wedge \sim q \wedge \sim r$ (4) $\sim p \wedge q \wedge r$,

    where p, q, and r are propositions. How many of these imply the truth of $(p \rightarrow q) \rightarrow r$?

    (A) 0   (B) 1   (C) 2   (D) 3   (E) 4

    Form II

    Consider the statements (1) p and r are true and q is false (2) r is true and p and q are false (3) p is true and q and r are false (4) q and r are true and p is false. How many of these imply the truth of the statement "r is implied by the statement that p implies q"?

    (A) 0   (B) 1   (C) 2   (D) 3   (E) 4

27. Six straight lines are drawn in a plane with no two parallel and no three concurrent. The number of regions into which they divide the plane is:

    (A) 16   (B) 20   (C) 22   (D) 24   (E) 26

28. Given the equation $3x^2 - 4x + k = 0$ with real roots. The value of k for which the product of the roots of the equation is a maximum is:

    (A) $\frac{16}{9}$ (B) $\frac{16}{3}$ (C) $\frac{4}{9}$ (D) $\frac{4}{3}$ (E) $-\frac{4}{3}$

29. A particle projected vertically upward reaches, at the end of t seconds, an elevation of s feet where $s = 160\,t - 16t^2$. The highest elevation is:

(A) 800   (B) 640   (C) 400   (D) 320   (E) 160

30. Let
$$F = \log \frac{1+x}{1-x}.$$
Form a new function G by replacing each x in F by
$$\frac{3x + x^3}{1 + 3x^2},$$
and simplify. The simplified expression G is equal to:

(A) $-F$ (B) F (C) $3F$ (D) F^3 (E) $F^3 - F$

Part 3

31. The number of solutions in positive integers of $2x + 3y = 763$ is:

(A) 255   (B) 254   (C) 128   (D) 127   (E) 0

32. The dimensions of a rectangle R are a and b, $a < b$. It is required to obtain a rectangle with dimensions x and y, $x < a$, $y < a$, so that its perimeter is one-third that of R, and its area is one-third that of R. The number of such (different) rectangles is:

(A) 0   (B) 1   (C) 2   (D) 4   (E) infinitely many

33. Given the line $y = \tfrac{3}{4}x + 6$ and a line L parallel to the given line and 4 units from it. A possible equation for L is:

(A) $y = \tfrac{3}{4}x + 1$ (B) $y = \tfrac{3}{4}x$ (C) $y = \tfrac{3}{4}x - \tfrac{2}{3}$
(D) $y = \tfrac{3}{4}x - 1$ (E) $y = \tfrac{3}{4}x + 2$

34. In triangle ABC, side $a = \sqrt{3}$, side $b = \sqrt{3}$, and side $c > 3$. Let x be the largest number such that the magnitude, in degrees, of the angle opposite side c exceeds x. Then x equals:

(A) 150   (B) 120   (C) 105   (D) 90   (E) 60

35. The lengths of the sides of a triangle are integers, and its area is also an integer. One side is 21 and the perimeter is 48. The shortest side is:

(A) 8   (B) 10   (C) 12   (D) 14   (E) 16

36. A person starting with 64 cents and making 6 bets, wins three times and loses three times, the wins and losses occurring in random order. The chance for a win is equal to the chance for a loss. If each wager is for half the money remaining at the time of the bet, then the final result is:

(A) a loss of 27¢   (B) a gain of 27¢   (C) a loss of 37¢
(D) neither a gain nor a loss
(E) a gain or a loss depending upon the order in which the wins and losses occur

37. Given points P_1, P_2, $\cdots$, P_7 on a straight line, in the order stated (not necessarily evenly spaced). Let P be an arbitrarily selected point on the line and let s be the sum of the undirected lengths

$$PP_1, \ PP_2, \ \cdots, \ PP_7.$$

Then s is smallest if and only if the point P is:

(A) midway between P_1 and P_7 (B) midway between P_2 and P_6
(C) midway between P_3 and P_5 (D) at P_4 (E) at P_1

38. Point F is taken on the extension of side AD of parallelogram $ABCD$. BF intersects diagonal AC at E and side DC at G. If $EF = 32$ and $GF = 24$, then BE equals:

(A) 4   (B) 8   (C) 10   (D) 12   (E) 16

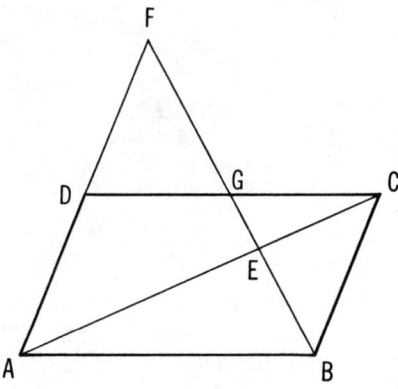

39. In triangle ABC lines CE and AD are drawn so that
$$\frac{CD}{DB} = \frac{3}{1} \quad \text{and} \quad \frac{AE}{EB} = \frac{3}{2}.$$ Let $r = \frac{CP}{PE}$,
where P is the intersection point of CE and AD. Then r equals:

(A) 3   (B) $\frac{3}{2}$ (C) 4   (D) 5   (E) $\frac{5}{2}$

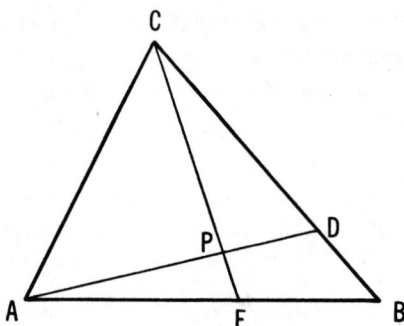

40. If x is a number satisfying the equation $\sqrt[3]{x+9} - \sqrt[3]{x-9} = 3$, then x^2 is between:

(A) 55 and 65   (B) 65 and 75   (C) 75 and 85
(D) 85 and 95   (E) 95 and 105

1964 Examination

Part 1

1. What is the value of $[\log_{10} (5 \log_{10} 100)]^2$?

   (A) $\log_{10} 50$ (B) 25   (C) 10   (D) 2   (E) 1

2. The graph of $x^2 - 4y^2 = 0$ is:

   (A) a parabola   (B) an ellipse   (C) a pair of straight lines
   (D) a point   (E) none of these

3. When a positive integer x is divided by a positive integer y, the quotient is u and the remainder is v, u and v integers. What is the remainder when $x + 2uy$ is divided by y?

   (A) 0   (B) $2u$ (C) $3u$ (D) v (E) $2v$

4. The expression

$$\frac{P+Q}{P-Q} - \frac{P-Q}{P+Q},$$

   where $P = x + y$ and $Q = x - y$, is equivalent to:

   (A) $\dfrac{x^2 - y^2}{xy}$ (B) $\dfrac{x^2 - y^2}{2xy}$ (C) 1   (D) $\dfrac{x^2 + y^2}{xy}$ (E) $\dfrac{x^2 + y^2}{2xy}$

5. If y varies directly as x, and if $y = 8$ when $x = 4$, the value of y when $x = -8$ is:

   (A) -16 (B) -4 (C) -2 (D) $4k$, $k = \pm 1, \pm 2, \cdots$
   (E) $16k$, $k = \pm 1, \pm 2, \cdots$

6. If x, $2x + 2$, $3x + 3$, $\cdots$ are in geometric progression, the fourth term is:

   (A) -27 (B) $-13\frac{1}{2}$ (C) 12   (D) $13\frac{1}{2}$ (E) 27

7. Let n be the number of real values of p for which the roots of
$$x^2 - px + p = 0$$
are equal. Then n equals:

(A) 0   (B) 1   (C) 2   (D) a finite number greater than 2
(E) an infinitely large number

8. The smaller root of the equation $(x - \frac{3}{4})(x - \frac{3}{4}) + (x - \frac{3}{4})(x - \frac{1}{2}) = 0$ is:

(A) $-\frac{3}{4}$ (B) $\frac{1}{2}$ (C) $\frac{5}{8}$ (D) $\frac{3}{4}$ (E) 1

9. A jobber buys an article at "$24 less $12\frac{1}{2}\%$". He then wishes to sell the article at a gain of $33\frac{1}{3}\%$ of his cost after allowing a 20% discount on his marked price. At what price, in dollars, should the article be marked?

(A) 25.20   (B) 30.00   (C) 33.60   (D) 40.00   (E) none of these

10. Given a square with side of length s. On a diagonal as base a triangle with three unequal sides is constructed so that its area equals that of the square. The length of the altitude drawn to the base is:

(A) $s\sqrt{2}$ (B) $s/\sqrt{2}$ (C) $2s$ (D) $2\sqrt{s}$ (E) $2/\sqrt{s}$

11. Given $2^x = 8^{y+1}$ and $9^y = 3^{x-9}$; the value of $x + y$ is:

(A) 18   (B) 21   (C) 24   (D) 27   (E) 30

12. Which of the following is the negation of the statement: For all x of a certain set, $x^2 > 0$?

(A) For all x, $x^2 < 0$ (B) For all x, $x^2 \leq 0$
(C) For no x, $x^2 > 0$ (D) For some x, $x^2 > 0$
(E) For some x, $x^2 \leq 0$

13. A circle is inscribed in a triangle with sides of lengths 8, 13, and 17, Let the segments of the side of length 8, made by a point of tangency. be r and s, with $r < s$. Then the ratio $r:s$ is:

(A) 1:3   (B) 2:5   (C) 1:2   (D) 2:3   (E) 3:4

14. A farmer bought 749 sheep. He sold 700 of them tor the price paid for the 749 sheep. The remaining 49 sheep were sold at the same price per

head as the other 700. Based on the cost, the percent gain on the entire transaction is:

(A) 6.5   (B) 6.75   (C) 7.0   (D) 7.5   (E) 8.0

15. A line through the point $(-a, 0)$ cuts from the second quadrant a triangular region with area T. The equation of the line is:

(A) $2Tx + a^2y + 2aT = 0$ (B) $2Tx - a^2y + 2aT = 0$
(C) $2Tx + a^2y - 2aT = 0$ (D) $2Tx - a^2y - 2aT = 0$
(E) none of these

16. Let $f(x) = x^2 + 3x + 2$ and let S be the set of integers $\{0, 1, 2, \cdots, 25\}$. The number of members s of S such that $f(s)$ has remainder zero when divided by 6 is:

(A) 25   (B) 22   (C) 21   (D) 18   (E) 17

17. Given the distinct points $P(x_1, y_1)$, $Q(x_2, y_2)$ and $R(x_1 + x_2, y_1 + y_2)$. Line segments are drawn connecting these points to each other and to the origin O. Of the three possibilities: (1) parallelogram (2) straight line (3) trapezoid, figure $OPRQ$, depending upon the location of the points $P, Q,$ and R, can be:

(A) (1) only   (B) (2) only   (C) (3) only   (D) (1) or (2) only
(E) none of the three possibilities mentioned

18. Let n be the number of pairs of values of b and c such that

$$3x + by + c = 0 \quad \text{and} \quad cx - 2y + 12 = 0$$

have the same graph. Then n is:

(A) 0   (B) 1   (C) 2   (D) finite but more than 2
(E) greater than any finite number

19. If $2x - 3y - z = 0$ and $x + 3y - 14z = 0$, $z \neq 0$, the numerical value of $(x^2 + 3xy)/(y^2 + z^2)$ is:

(A) 7   (B) 2   (C) 0   (D) $-20/17$ (E) -2

20. The sum of the numerical coefficients of all the terms in the expansion of $(x - 2y)^{18}$ is:

(A) 0   (B) 1   (C) 19   (D) -1 (E) -19

Part 2

21. If $\log_{b^2} x + \log_{x^2} b = 1$, $b > 0$, $b \neq 1$, $x \neq 1$, then x equals:

   (A) $1/b^2$ (B) $1/b$ (C) b^2 (D) b (E) $\sqrt{b}$

22. Given parallelogram $ABCD$ with E the midpoint of diagonal BD. Point E is connected to a point F in DA so that $DF = \frac{1}{3}DA$. What is the ratio of the area of triangle DFE to the area of quadrilateral $ABEF$?

   (A) 1:2   (B) 1:3   (C) 1:5   (D) 1.6   (E) 1:7

23. Two numbers are such that their difference, their sum, and their product are to one another as 1:7:24. The product of the two numbers is:

   (A) 6   (B) 12   (C) 24   (D) 48   (E) 96

24. Let $y = (x-a)^2 + (x-b)^2$, a, b constants. For what value of x is y a minimum?

   (A) $\dfrac{a+b}{2}$ (B) $a+b$ (C) $\sqrt{ab}$ (D) $\sqrt{\dfrac{a^2+b^2}{2}}$ (E) $\dfrac{a+b}{2ab}$

25. The set of values of m for which $x^2 + 3xy + x + my - m$ has two factors, with integer coefficients, which are linear in x and y, is precisely:

   (A) 0, 12, -12 (B) 0, 12   (C) 12, -12 (D) 12   (E) 0

26. In a ten-mile race First beats Second by 2 miles and First beats Third by 4 miles. If the runners maintain constant speeds throughout the race, by how many miles does Second beat Third?

   (A) 2   (B) $2\frac{1}{4}$ (C) $2\frac{1}{2}$ (D) $2\frac{3}{4}$ (E) 3

27. If x is a real number and $|x-4| + |x-3| < a$ where $a > 0$, then:

   (A) $0 < a < .01$ (B) $.01 < a < 1$ (C) $0 < a < 1$
   (D) $0 < a \leq 1$ (E) $a > 1$

28. The sum of n terms of an arithmetic progression is 153, and the common difference is 2. If the first term is an integer, and $n > 1$, then the number of possible values for n is:

   (A) 2   (B) 3   (C) 4   (D) 5   (E) 6

PROBLEMS: 1964 EXAMINATION   35

29. In this figure $\angle RFS = \angle FDR$, $FD = 4$ inches, $DR = 6$ inches, $FR = 5$ inches, $FS = 7\tfrac{1}{2}$ inches. The length of RS, in inches, is:

(A) undetermined (B) 4 (C) $5\tfrac{1}{2}$ (D) 6 (E) $6\tfrac{1}{4}$

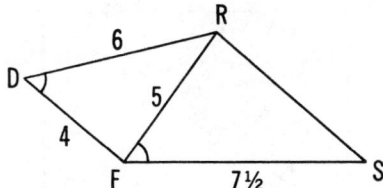

30. The larger root minus the smaller root of the equation
$$(7 + 4\sqrt{3})x^2 + (2 + \sqrt{3})x - 2 = 0$$
is

(A) $-2 + 3\sqrt{3}$ (B) $2 - \sqrt{3}$ (C) $6 + 3\sqrt{3}$
(D) $6 - 3\sqrt{3}$ (E) $3\sqrt{3} + 2$

Part 3

31. Let
$$f(n) = \frac{5 + 3\sqrt{5}}{10}\left(\frac{1 + \sqrt{5}}{2}\right)^n + \frac{5 - 3\sqrt{5}}{10}\left(\frac{1 - \sqrt{5}}{2}\right)^n.$$
Then $f(n + 1) - f(n - 1)$, expressed in terms of $f(n)$, equals:

(A) $\tfrac{1}{2}f(n)$ (B) $f(n)$ (C) $2f(n) + 1$ (D) $f^2(n)$
(E) $\tfrac{1}{2}(f^2(n) - 1)$

32. If $\dfrac{a + b}{b + c} = \dfrac{c + d}{d + a}$, then:

(A) a must equal c (B) $a + b + c + d$ must equal zero
(C) either $a = c$ or $a + b + c + d = 0$, or both
(D) $a + b + c + d \neq 0$ if $a = c$
(E) $a(b + c + d) = c(a + b + d)$

33. P is a point interior to rectangle $ABCD$ and such that $PA = 3$ inches, $PD = 4$ inches, and $PC = 5$ inches. Then PB, in inches, equals:

(A) $2\sqrt{3}$ (B) $3\sqrt{2}$ (C) $3\sqrt{3}$ (D) $4\sqrt{2}$ (E) 2

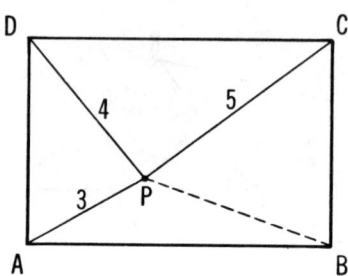

34. If n is a multiple of 4, the sum $s = 1 + 2i + 3i^2 + \cdots + (n+1)i^n$, where $i = \sqrt{-1}$, equals:

(A) $1 + i$ (B) $\tfrac{1}{2}(n+2)$ (C) $\tfrac{1}{2}(n+2-ni)$
(D) $\tfrac{1}{2}[(n+1)(1-i)+2]$ (E) $\tfrac{1}{8}(n^2+8-4ni)$

35. The sides of a triangle are of lengths 13, 14, and 15. The altitudes of the triangle meet at point H. If AD is the altitude to side of length 14, the ratio $HD:HA$ is:

(A) $3:11$ (B) $5:11$ (C) $1:2$ (D) $2:3$ (E) $25:33$

36. In this figure the radius of the circle is equal to the altitude of the equilateral triangle ABC. The circle is made to roll along the side AB, remaining tangent to it at a variable point T and intersecting sides AC and BC in variable points M and N, respectively. Let n be the number of degrees in arc MTN. Then n, for all permissible positions of the circle:

(A) varies from 30° to 90°

(B) varies from 30° to 60°

(C) varies from 60° to 90°

(D) remains constant at 30°

(E) remains constant at 60°

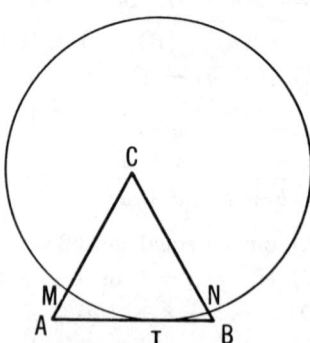

37. Given two positive numbers a, b such that $a < b$. Let A.M. be their arithmetic mean and let G.M. be their positive geometric mean. Then A.M. minus G.M. is always less than:

(A) $\dfrac{(b+a)^2}{ab}$ (B) $\dfrac{(b+a)^2}{8b}$ (C) $\dfrac{(b-a)^2}{ab}$

(D) $\dfrac{(b-a)^2}{8a}$ (E) $\dfrac{(b-a)^2}{8b}$

38. The sides PQ and PR of triangle PQR are respectively of lengths 4 inches and 7 inches. The median PM is $3\frac{1}{2}$ inches. Then QR, in inches, is:

(A) 6   (B) 7   (C) 8   (D) 9   (E) 10

39. The magnitudes of the sides of triangle ABC are a, b, and c, as shown, with $c \leq b \leq a$. Through interior point P and the vertices A, B, C, lines are drawn meeting the opposite sides in A', B', C', respectively. Let $s = AA' + BB' + CC'$. Then, for all positions of point P, s is less than:

(A) $2a + b$ (B) $2a + c$ (C) $2b + c$ (D) $a + 2b$
(E) $a + b + c$

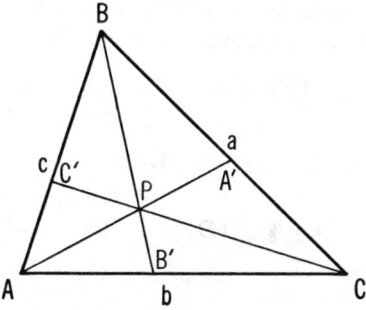

40. A watch loses $2\frac{1}{2}$ minutes per day. It is set right at 1 P.M. on March 15. Let n be the positive correction, in minutes, to be added to the time shown by the watch at a given time. When the watch shows 9 A.M. on March 21, n equals:

(A) $14\frac{14}{23}$ (B) $14\frac{1}{14}$ (C) $13\frac{101}{115}$ (D) $13\frac{83}{115}$ (E) $13\frac{13}{23}$

1965 Examination

Part 1

1. The number of real values of x satisfying the equation $2^{2x^2-7x+5} = 1$ is:

   (A) 0 (B) 1 (C) 2 (D) 4 (E) more than 4

2. A regular hexagon is inscribed in a circle. The ratio of the length of a side of the hexagon to the length of the shorter of the arcs intercepted by the side, is:

   (A) 1:1 (B) 1:6 (C) 1:π (D) 3:π (E) 6:π

3. The expression $(81)^{-(2^{-2})}$ has the same value as:

   (A) $\frac{1}{81}$ (B) $\frac{1}{3}$ (C) 3 (D) 81 (E) 81^4

4. Line l_2 intersects line l_1 and line l_3 is parallel to l_1. The three lines are distinct and lie in a plane. The number of points equidistant from all three lines is:

   (A) 0 (B) 1 (C) 2 (D) 4 (E) 8

5. When the repeating decimal $0.363636\cdots$ is written in simplest fractional form, the sum of the numerator and denominator is:

   (A) 15 (B) 45 (C) 114 (D) 135 (E) 150

6. If $10^{\log_{10} 9} = 8x + 5$, then x equals:

   (A) 0 (B) $\frac{1}{2}$ (C) $\frac{5}{8}$ (D) $\frac{9}{8}$ (E) $\dfrac{2\log_{10} 3 - 5}{8}$

7. The sum of the reciprocals of the roots of the equation $ax^2 + bx + c = 0$ is:

   (A) $\dfrac{1}{a} + \dfrac{1}{b}$ (B) $-\dfrac{c}{b}$ (C) $\dfrac{b}{c}$ (D) $-\dfrac{a}{b}$ (E) $-\dfrac{b}{c}$

8. One side of a given triangle is 18 inches. Inside the triangle a line segment is drawn parallel to this side forming a trapezoid whose area is one-third

of that of the triangle. The length of this segment, in inches, is:

(A) $6\sqrt{6}$ (B) $9\sqrt{2}$ (C) 12 (D) $6\sqrt{3}$ (E) 9

9. The vertex of the parabola $y = x^2 - 8x + c$ will be a point on the x-axis if the value of c is:

(A) -16 (B) -4 (C) 4 (D) 8 (E) 16

10. The statement $x^2 - x - 6 < 0$ is equivalent to the statement:

(A) $-2 < x < 3$ (B) $x > -2$ (C) $x < 3$
(D) $x > 3$ and $x < -2$ (E) $x > 3$ or $x < -2$

11. Consider the statements:

I: $(\sqrt{-4})(\sqrt{-16}) = \sqrt{(-4)(-16)}$,

II: $\sqrt{(-4)(-16)} = \sqrt{64}$,

and

III: $\sqrt{64} = 8$.

Of these the following are *incorrect*

(A) none (B) I only (C) II only (D) III only
(E) I and III only

12. A rhombus is inscribed in triangle ABC in such a way that one of its vertices is A and two of its sides lie along AB and AC. If $AC = 6$ inches, $AB = 12$ inches, and $BC = 8$ inches, the side of the rhombus, in inches, is:

(A) 2 (B) 3 (C) $3\frac{1}{2}$ (D) 4 (E) 5

13. Let n be the number of number-pairs (x, y) which satisfy $5y - 3x = 15$ and $x^2 + y^2 \leq 16$. Then n is:

(A) 0 (B) 1 (C) 2 (D) more than two, but finite
(E) greater than any finite number

14. The sum of the numerical coefficients in the complete expansion of $(x^2 - 2xy + y^2)^7$ is:

(A) 0 (B) 7 (C) 14 (D) 128 (E) 128^2

15. The symbol 25_b represents a two-digit number in the base b. If the number 52_b is double the number 25_b, then b is:

(A) 7   (B) 8   (C) 9   (D) 11   (E) 12

16. Let line AC be perpendicular to line CE. Connect A to the midpoint D of CE, and connect E to the midpoint B of AC. If AD and EB intersect in point F, and $BC = CD = 15$ inches, then the area of triangle DFE, in square inches, is:

(A) 50   (B) $50\sqrt{2}$ (C) 75   (D) $\tfrac{15}{2}\sqrt{105}$ (E) 100

17. Given the true statement: The picnic on Sunday will not be held only if the weather is not fair. We can then conclude that:

(A) If the picnic is held, Sunday's weather is undoubtedly fair.

(B) If the picnic is not held, Sunday's weather is possibly unfair.

(C) If it is not fair Sunday, the picnic will not be held.

(D) If it is fair Sunday, the picnic may be held.

(E) If it is fair Sunday, the picnic will be held.

18. If $1 - y$ is used as an approximation to the value of

$$\frac{1}{1+y}, \quad |y| < 1,$$

the ratio of the error made to the correct value is:

(A) y (B) y^2 (C) $\dfrac{1}{1+y}$ (D) $\dfrac{y}{1+y}$ (E) $\dfrac{y^2}{1+y}$

19. If $x^4 + 4x^3 + 6px^2 + 4qx + r$ is exactly divisible by $x^3 + 3x^2 + 9x + 3$, the value of $(p+q)r$ is:

(A) -18 (B) 12   (C) 15   (D) 27   (E) 45

20. For every n the sum S_n of n terms of an arithmetic progression is $2n + 3n^2$. The rth term is:

(A) $3r^2$ (B) $3r^2 + 2r$ (C) $6r - 1$ (D) $5r + 5$ (E) $6r + 2$

Part 2

21. It is possible to choose $x > \frac{2}{3}$ in such a way that the value of
$$\log_{10}(x^2 + 3) - 2\log_{10} x$$
is:

(A) negative (B) zero (C) one
(D) smaller than any positive number that might be specified
(E) greater than any positive number that might be specified

22. If $a_2 \neq 0$ and r and s are the roots of $a_0 + a_1 x + a_2 x^2 = 0$, then the equality $a_0 + a_1 x + a_2 x^2 = a_0(1 - x/r)(1 - x/s)$ holds:

(A) for all values of x, $a_0 \neq 0$ (B) for all values of x
(C) only when $x = 0$ (D) only when $x = r$ or $x = s$
(E) only when $x = r$ or $x = s$, $a_0 \neq 0$

23. If we write $|x^2 - 4| < N$ for all x such that $|x - 2| < 0.01$, the smallest value we can use for N is:

(A) .0301 (B) .0349 (C) .0399 (D) .0401 (E) .0499

24. Given the sequence $10^{1/11}, 10^{2/11}, 10^{3/11}, \cdots, 10^{n/11}$, the smallest value of n such that the product of the first n members of this sequence exceeds 100,000 is:

(A) 7 (B) 8 (C) 9 (D) 10 (E) 11

25. Let $ABCD$ be a quadrilateral with AB extended to E so that $AB = BE$. Lines AC and CE are drawn to form angle ACE. For this angle to be a right angle it is necessary that quadrilateral $ABCD$ have:

(A) all angles equal (B) all sides equal (C) two pairs of equal sides
(D) one pair of equal sides (E) one pair of equal angles

26. For the numbers a, b, c, d, e, define m to be the arithmetic mean of all five numbers; k to be the arithmetic mean of a and b; l to be the arithmetic mean of $c, d,$ and e; and p to be the arithmetic mean of k and l. Then, no matter how a, b, c, d, e are chosen, we shall always have:

(A) $m = p$ (B) $m \geq p$ (C) $m > p$
(D) $m < p$ (E) none of these

27. When $y^2 + my + 2$ is divided by $y - 1$ the quotient is $f(y)$ and the remainder is R_1. When $y^2 + my + 2$ is divided by $y + 1$ the quotient is $g(y)$ and the remainder is R_2. If $R_1 = R_2$ then m is:

(A) 0   (B) 1   (C) 2   (D) −1   (E) an undetermined constant

28. An escalator (moving staircase) of n uniform steps visible at all times descends at constant speed. Two boys, A and Z, walk down the escalator steadily as it moves, A negotiating twice as many escalator steps per minute as Z. A reaches the bottom after taking 27 steps while Z reaches the bottom after taking 18 steps. Then n is:

(A) 63   (B) 54   (C) 45   (D) 36   (E) 30

29. Of 28 students taking at least one subject, the number taking Mathematics and English only equals the number taking Mathematics only. No student takes English only or History only, and six students take Mathematics and History, but no English. The number taking English and History only is five times the number taking all three subjects. If the number taking all three subjects is even and non-zero, the number taking English and Mathematics only is:

(A) 5   (B) 6   (C) 7   (D) 8   (E) 9

30. Let BC of right triangle ABC be the diameter of a circle intersecting hypotenuse AB in D. At D a tangent is drawn cutting leg CA in F. This information is *not* sufficient to prove that

(A) DF bisects CA (B) DF bisects $\angle CDA$
(C) $DF = FA$ (D) $\angle A = \angle BCD$ (E) $\angle CFD = 2 \angle A$

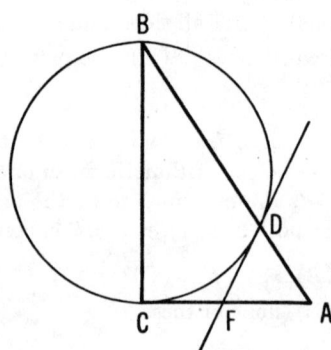

Part 3

31. The number of real values of x satisfying the equality

$$(\log_a x)(\log_b x) = \log_a b,$$

where a and b are positive constants different from 1, is

(A) 0 (B) 1 (C) 2 (D) an integer greater than 2
(E) not finite

32. An article costing C dollars is sold for $100 at a loss of x percent of the selling price. It is then resold at a profit of x percent of the new selling price S'. If the difference between S' and C is $1\frac{1}{9}$ dollars, then x is:

(A) undetermined (B) $\frac{80}{9}$ (C) 10 (D) $\frac{95}{9}$ (E) $\frac{100}{9}$

33. If the number 15!, that is, $15 \cdot 14 \cdot 13 \cdots 1$, ends with k zeros when given to the base 12 and ends with h zeros when given to the base 10, then $k + h$ equals:

(A) 5 (B) 6 (C) 7 (D) 8 (E) 9

34. For $x \geq 0$, the smallest value of $\dfrac{4x^2 + 8x + 13}{6(1 + x)}$ is:

(A) 1 (B) 2 (C) $\frac{25}{12}$ (D) $\frac{13}{6}$ (E) $\frac{34}{5}$

35. The length of a rectangle is 5 inches and its width is less than 4 inches. The rectangle is folded so that two diagonally opposite vertices coincide. If the length of the crease is $\sqrt{6}$, then the width is:

(A) $\sqrt{2}$ (B) $\sqrt{3}$ (C) 2 (D) $\sqrt{5}$ (E) $\sqrt{11/2}$

36. Given distinct straight lines OA and OB. From a point in OA a perpendicular is drawn to OB; from the foot of this perpendicular a line is drawn perpendicular to OA. From the foot of this second perpendicular a line is drawn perpendicular to OB; and so on indefinitely. The lengths of the first and second perpendiculars are a and b, respectively. Then the sum of the lengths of the perpendiculars approaches a limit as the number of perpendiculars grows beyond all bounds. This limit is:

(A) $\dfrac{b}{a-b}$ (B) $\dfrac{a}{a-b}$ (C) $\dfrac{ab}{a-b}$ (D) $\dfrac{b^2}{a-b}$ (E) $\dfrac{a^2}{a-b}$

37. Point E is selected on side AB of triangle ABC in such a way that $AE:EB = 1:3$ and point D is selected on side BC so that $CD:DB = 1:2$. The point of intersection of AD and CE is F. Then

$$\frac{EF}{FC} + \frac{AF}{FD} \quad \text{is:}$$

(A) $\frac{4}{5}$ (B) $\frac{5}{4}$ (C) $\frac{3}{2}$ (D) 2 (E) $\frac{5}{2}$

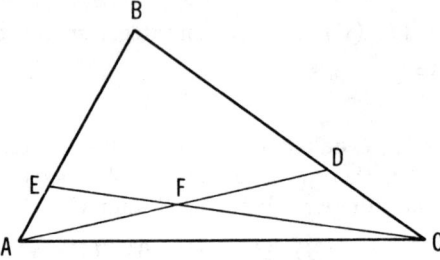

38. A takes m times as long to do a piece of work as B and C together; B takes n times as long as C and A together; and C takes x times as long as A and B together. Then x, in terms of m and n, is:

(A) $\dfrac{2mn}{m+n}$ (B) $\dfrac{1}{2(m+n)}$ (C) $\dfrac{1}{m+n-mn}$

(D) $\dfrac{1-mn}{m+n+2mn}$ (E) $\dfrac{m+n+2}{mn-1}$

39. A foreman noticed an inspector checking a $3''$-hole with a $2''$-plug and a $1''$-plug and suggested that two more gauges be inserted to be sure that the fit was snug. If the new gauges are alike, then the diameter, d, of each, to the nearest hundredth of an inch, is:

(A) .87 (B) .86 (C) .83 (D) .75 (E) .71

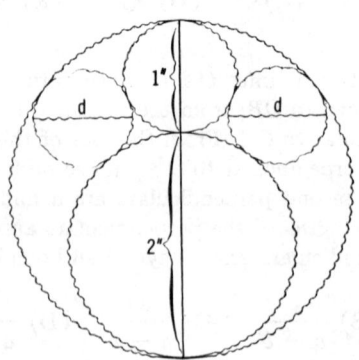

40. Let n be the number of integer values of x such that
$$P = x^4 + 6x^3 + 11x^2 + 3x + 31$$
is the square of an integer. Then n is:

(A) 4   (B) 3   (C) 2   (D) 1   (E) 0

II

Answer Keys

1961 Answers

1. C	11. C	21. D	31. D
2. E	12. A	22. C	32. C
3. E	13. E	23. B	33. B
4. B	14. E	24. A	34. A
5. C	15. B	25. A	35. D
6. E	16. E	26. C	36. B
7. A	17. D	27. D	37. C
8. B	18. A	28. D	38. A
9. C	19. B	29. B	39. B
10. D	20. A	30. D	40. A

1962 Answers

1. D	11. B	21. E	31. C
2. A	12. C	22. D	32. E
3. B	13. B	23. E	33. A
4. B	14. B	24. A	34. E
5. C	15. D	25. C	35. B
6. D	16. C	26. E	36. C
7. C	17. B	27. E	37. D
8. D	18. A	28. D	38. B
9. B	19. C	29. A	39. C
10. A	20. A	30. D	40. B

1963 Answers

1. D	11. B	21. E	31. D
2. A	12. E	22. D	32. A
3. E	13. E	23. B	33. A
4. D	14. A	24. B	34. B
5. E	15. C	25. A	35. B
6. B	16. C	26. E	36. C
7. A	17. A	27. C	37. D
8. D	18. A	28. D	38. E
9. C	19. B	29. C	39. D
10. B	20. D	30. C	40. C

1964 Answers

1. E	11. D	21. D	31. B
2. C	12. E	22. C	32. C
3. D	13. A	23. D	33. B
4. A	14. C	24. A	34. C
5. A	15. B	25. B	35. B
6. B	16. E	26. C	36. E
7. C	17. D	27. E	37. D
8. C	18. C	28. D	38. D
9. E	19. A	29. E	39. A
10. A	20. B	30. D	40. A

1965 Answers

1. C	6. B	11. B	16. C	21. D	26. E	31. C	36. E
2. D	7. E	12. D	17. E	22. A	27. A	32. C	27. C
3. B	8. A	13. E	18. B	23. D	28. B	33. D	38. E
4. C	9. E	14. A	19. C	24. E	29. A	34. B	39. B
5. A	10. A	15. B	20. C	25. D	30. B	35. D	40. D

III

Solutions*

1961 Solutions

Part 1

1. (C) $(-\frac{1}{125})^{-2/3} = (-125)^{2/3} = ((-125)^{1/3})^2 = (-5)^2 = 25.$

2. (E) D (yards) $= R$ (yards per minute) $\times T$ (minutes);
$$D = \frac{(1/3)\cdot(a/6)}{r/60}\cdot 3 = \frac{10a}{r}.$$

3. (E) Let m_1 be the slope of the line $2y + x + 3 = 0$, and m_2 the slope of the line $3y + ax + 2 = 0$. For perpendicularity, $m_1 m_2 = -1$;
$$\therefore \left(-\frac{1}{2}\right)\left(-\frac{a}{3}\right) = -1, \qquad \therefore a = -6.$$

4. (B) u is not closed under addition since, for example, $1 + 4 = 5$ and 5, not being a perfect square of an integer, is not a member of u. Similarly, u is not closed under division or positive integral root extraction. But u is closed under multiplication because, if m^2 and n^2 are elements of u, $m^2 \cdot n^2 = (mn)^2$ is a member of u (u contains the squares of all positive integers and mn is an integer).

---

* The letter following the problem number refers to the correct choice of the five listed in the examination.

5. (C) Consider the binomial expansion
$$(a + 1)^4 = a^4 + 4a^3 + 6a^2 + 4a + 1.$$
Let $x - 1 = a$. Then $S = (x - 1 + 1)^4 = x^4$.

6. (E) $\log 8 \div \log \frac{1}{8} = \log 8 \div (\log 1 - \log 8)$
$$= \log 8 \div (-\log 8) = -1.$$

7. (A) Since $(r + s)^6 = r^6 + 6r^5 s + 15r^4 s^2 + \cdots$, we have, setting
$$r = \frac{a}{\sqrt{x}} \quad \text{and} \quad s = -\frac{\sqrt{x}}{a^2}$$
in the third term,
$$15\left(\frac{a}{\sqrt{x}}\right)^4 \left(-\frac{\sqrt{x}}{a^2}\right)^2 = \frac{15}{x}.$$

OR

Use the formula
$$\frac{n(n-1) \cdots (n-k+1)}{1 \cdot 2 \cdots k}(r)^{n-k}(s)^k$$
for the $(k+1)$th term of $(r+s)^n$. For $n = 6$ and $k = 2$, we have
$$\frac{6 \cdot 5}{1 \cdot 2}\left(\frac{a}{\sqrt{x}}\right)^4 \left(-\frac{\sqrt{x}}{a^2}\right)^2 = \frac{15}{x}.$$

8. (B) Since $B + C_2 = 90°$ and $A + C_1 = 90°$, $B + C_2 = A + C_1$.
$\therefore C_1 - C_2 = B - A$.

9. (C) $r = (2a)^{2b} = ((2a)^2)^b = (4a^2)^b$; also $r = a^b \cdot x^b = (ax)^b$.
$\therefore (ax)^b = (4a^2)^b$, $ax = 4a^2$, $x = 4a$.

10. (D) $BE^2 = BD^2 + DE^2$, $BD = 6$, $DE = \frac{1}{2}DA = \frac{1}{2} \cdot 6\sqrt{3} = 3\sqrt{3}$.
$\therefore BE^2 = 36 + 27$ and $BE = (63)^{1/2}$.

11. (C) p (perimeter of triangle APR) $= AP + PQ + RQ + RA$, and $PQ = PB$ and $RQ = RC$.
$\therefore p = AP + PB + RC + RA = AB + AC = 20 + 20 = 40$.

SOLUTIONS: 1961 EXAMINATION 51

12. (A) In a geometric progression, each term is r times the preceding one. So in this case, $r = 2^{1/3}/2^{1/2} = 2^{1/3-1/2} = 2^{-1/6}$. Thus, the fourth term is $r \cdot 2^{1/6} = 2^{-1/6} 2^{1/6} = 2^0 = 1$.

13. (E) $(t^4 + t^2)^{1/2} = [t^2(t^2 + 1)]^{1/2} = (t^2)^{1/2}(t^2 + 1)^{1/2} = |t|(1 + t^2)^{1/2}$.

14. (E) Let the diagonals have lengths d and $2d$, and let the side have length s. Then $\frac{1}{2}d \cdot 2d = K$ or $d^2 = K$. Since

$$s^2 = d^2 + \left(\frac{d}{2}\right)^2 = K + \frac{K}{4} = \frac{5K}{4}, \qquad s = \tfrac{1}{2}(5K)^{1/2}.$$

15. (B) Since x men working x^2 hours produce x articles, one man working x^2 hours produces one article. Therefore, one man produces $1/x^2$ part of one article in one hour. Let n be the number of articles produced by y men working y hours a day for each of y days. Then

$$n = y \cdot y \cdot y \cdot \frac{1}{x^2} = \frac{y^3}{x^2}.$$

16. (E) Let d be the amount taken from the base b. Then

$$\tfrac{1}{2}(b - d)(h + m) = \tfrac{1}{2} \cdot \tfrac{1}{2} bh.$$

Solving for d, we obtain

$$d = \frac{b(2m + h)}{2(h + m)}.$$

17. (D) We are given that 1000 m.u. $- 340$ m.u. $= 440$ m.u. (base r). Therefore

$$1 \cdot r^3 + 0 \cdot r^2 + 0 \cdot r + 0 - (3r^2 + 4r + 0) = 4r^2 + 4r + 0.$$

So

$$r^3 - 7r^2 - 8r = 0, \qquad r^2 - 7r - 8 = 0 \qquad r = 8.$$

OR

We have

$$\begin{array}{r} 440 \\ +\ 340 \\ \hline 1000; \end{array}$$

since $4 + 4$ terminates in 0 in this system, the base is 8, that is, $r = 8$.

18. (A) Let P be the original population. Then,

after 1 year the population is $1.25P$
after 2 years the population is $(1.25)(1.25)P$
after 3 years the population is $(1.25)(1.25)(.75)P$
after 4 years the population is

$$(1.25)(1.25)(.75)(.75)P = \tfrac{225}{256}P.$$

The net percent change is

$$\frac{\tfrac{225}{256}P - P}{P} \times 100 = \frac{0.88P - P}{P} \times 100 = -12.$$

19. (B) Writing $2 \log x$ as $\log x^2$, we obtain the abscissas of the intersection points by solving the equation $x^2 = 2x$. Since the possible values of x are restricted to positive numbers, there is but one root, namely 2. Hence, the graphs intersect only in $(2, \log 4)$.

20. (A) The set of points satisfying the inequalities $y > 2x$ and $y > 4 - x$ is the hatched region shown.

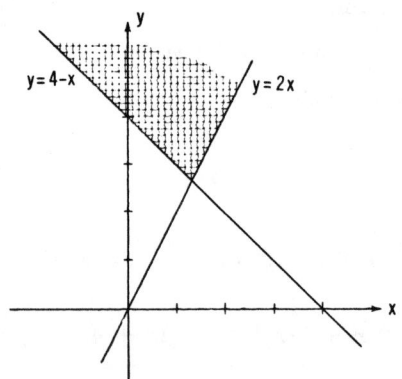

Part 2

21. (D) Area $(\triangle MNE) = \tfrac{1}{2}$ Area $(\triangle MAE)$
$= \tfrac{1}{2} \cdot \tfrac{1}{3}$ Area $(\triangle CAE)$
$= \tfrac{1}{2} \cdot \tfrac{1}{3} \cdot \tfrac{1}{2}$ Area $(\triangle ABC)$.
$\therefore \quad k = \tfrac{1}{12}.$

SOLUTIONS: 1961 EXAMINATION   53

22. (C) If $x - 3$ is a factor of $3x^3 - 9x^2 + kx - 12 = P(x)$, then $x = 3$ is a root of $P(x) = 0$, i.e. $P(3) = 0$; but

$$P(3) = 3 \cdot 3^3 - 9 \cdot 3^2 + 3k - 12 = 3k - 12 = 0$$

implies $k = 4$. Divide $3x^3 - 9x^2 + 4x - 12$ by $x - 3$ and obtain $3x^2 + 4$.

OR

The remainder, when $P(x)$ is divided by $x - 3$, is zero if and only if $12 = 3k$, so $k = 4$ and the quotient is $3x^2 + k = 3x^2 + 4$.

23. (B) Since P divides AB in the ratio $2:3$,

$$AP:AB = 2:5 \quad \text{and} \quad AQ:AB = 3:7.$$

$$\frac{AQ - AP}{AB} = \frac{1}{35}.$$

Since $PQ = AQ - AP = 2$, $AB = 70$.

24. (A) Let the prices, arranged in order from left to right be P, $P + 2$, $P + 4$, $\cdots$, $P + 58$, $P + 60$. The price of the middle book is $P + 30$. Then either

$$P + 30 + P + 32 = P + 60 \quad \text{or} \quad P + 30 + P + 28 = P + 60.$$

The first equation yields a negative value for P, an impossibility. The second equation leads to $P = 2$, so that (A) is the correct choice.

25. (A) Represent the magnitude of angle B by m. Then, in order, we obtain angle $QPB = m$, angle $AQP = 2m$, angle $QAP = 2m$, angle $QPA = 180 - 4m$, angle $APC = 3m$, angle $ACP = 3m$. Since angle BCA = angle $BAC = 3m$, the sum of the angles in $\triangle ABC$ is $m + 3m + 3m = 7m = 180°$. $\therefore m = 25\frac{5}{7}°$.

26. (C) $200 = \frac{50}{2}(a + a + 49d)$ and $2700 = \frac{50}{2}((a + 50d) + (a + 99d))$. Solve these equations simultaneously to obtain $a = -20.5$.

27. (D) The minimum number of sides in a polygon is 3, so the smallest possible value for the angle x is $60°$. The next largest possible value for x is $90°$ and belongs to a four-sided polygon. Each angle of a (convex) polygon is less than $180°$, so $kx < 180°$. Since k is an integer greater than 1, the pair $x = 60°$, $k = 2$ furnishes a

solution: $x = 60°$, $kx = 120° < 180°$; P_1 is a triangle, P_2 a hexagon. But if $x > 60°$ or if $k > 2$, $kx \geq 180°$; so there are no other solutions.

28. (D) The final digits for $(2137)^n$ are respectively 1 if $n = 0$, 7 if $n = 1$, 9 if $n = 2$, and 3 if $n = 3$. For larger values of n these digits repeat in cycles of four.

$$\therefore (2137)^{753} = (2137)^{4 \cdot 188 + 1} = (2137^4)^{188} \cdot (2137)^1.$$

The final digit of $(2137^4)^{188}$ is 1 and the final digit of $(2137)^1$ is 7. The product yields a final digit of $1 \cdot 7 = 7$.

29. (B) We have $r + s = -b/a$ and $rs = c/a$. The equation with roots $ar + b$ and $as + b$ is $(x - (ar + b))(x - (as + b)) = 0$.

$$\therefore x^2 - (a(r + s) + 2b)x + a^2 rs + ab(r + s) + b^2 = 0.$$

$$\therefore x^2 - \left(a\left(-\frac{b}{a}\right) + 2b\right)x + a^2\left(\frac{c}{a}\right) + ab\left(-\frac{b}{a}\right) + b^2 = 0.$$

$$\therefore x^2 - bx + ac = 0.$$

30. (D) Let $\log_5 12 = x$; then $5^x = 12$, so $x \log_{10} 5 = \log_{10} 12$.

$$\therefore x = \frac{\log_{10} 12}{\log_{10} 5} = \frac{2 \log_{10} 2 + \log_{10} 3}{\log_{10} 10 - \log_{10} 2} = \frac{2a + b}{1 - a}.$$

Part 3

31. (D) Draw PA' so that $\angle BPC = \angle A'PC$; then $\triangle ACP \cong \triangle A'CP$ (a s a) and $AC = A'C$, $PA = PA'$. Since PC bisects $\angle BPA'$ in $\triangle BPA'$,

$$\frac{BC}{CA'} = \frac{PB}{PA'} \quad \text{or} \quad \frac{BC}{CA} = \frac{PB}{PA} = \frac{4}{3}.$$

$AB = PB - PA$ since A is between P and B.

$$\frac{AB}{PA} = \frac{PB}{PA} - \frac{PA}{PA} = \frac{4}{3} - 1 = \frac{1}{3}, \quad \text{so} \quad \frac{PA}{AB} = 3.$$

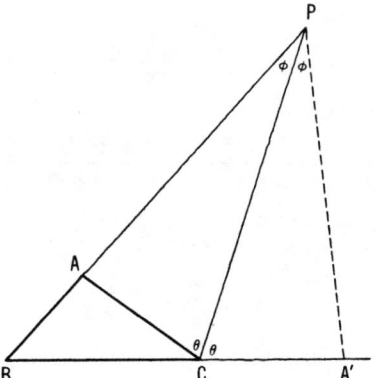

32. (C) The area of the regular polygon is $\frac{1}{2}ap$, where

$$a = R \cos \frac{180°}{n} \quad \text{and} \quad p = ns = n \cdot 2R \sin \frac{180°}{n}.$$

$$\therefore 3R^2 = \tfrac{1}{2} R \cos \frac{180°}{n} \cdot 2nR \sin \frac{180°}{n}.$$

$$\therefore \frac{6}{n} = 2 \sin \frac{180°}{n} \cos \frac{180°}{n} = \sin \frac{360°}{n},$$

where n is a positive integer equal to or greater than 3. Of the possible angles the only one whose sine is a rational number is 30°. $\therefore n = 12$. To check, note that $\frac{6}{12} = \frac{1}{2} = \sin \frac{360}{12} = \sin 30°$.

OR

The area of the polygon is n times the area of one triangle whose vertices are the center of the circle and two consecutive vertices of the polygon;

$$\therefore 3R^2 = n \cdot \tfrac{1}{2} R^2 \sin \frac{360°}{n} \quad \text{or, as before,} \quad \frac{6}{n} = \sin \frac{360°}{n}.$$

33. (B) $\quad 2^{2x} - 3^{2y} = (2^x + 3^y)(2^x - 3^y) = 11 \cdot 5 = 55 \cdot 1$

$\therefore 2^x + 3^y = 11 \qquad 2^x + 3^y = 55$

$ 2^x - 3^y = 5 \qquad 2^x - 3^y = 1$

$ 2 \cdot 2^x = 16 \qquad$ This system does not

$ x = 3 \qquad\quad$ yield integral values

$ y = 1 \qquad\quad$ for x and y.

34. (A) Let
$$y = \frac{2x+3}{x+2}.$$
$$\therefore y = \frac{2x+4-1}{x+2} = \frac{2(x+2)-1}{x+2} = 2 - \frac{1}{x+2}.$$

From the form
$$y = 2 - \frac{1}{x+2}$$
we see that y increases as x increases for all $x \geq 0$. Thus the smallest value of y is obtained when $x = 0$. $\therefore m = \frac{3}{2}$ and m is in S. As x continues to increase, y approaches 2 but never becomes equal to 2. $\therefore M = 2$ and M is not in S.

35. (D) $695 = a_1 + a_2(2 \cdot 1) + a_3(3 \cdot 2 \cdot 1)$
$\qquad + a_4(4 \cdot 3 \cdot 2 \cdot 1) + a_5(5 \cdot 4 \cdot 3 \cdot 2 \cdot 1),$

that is, $695 = a_1 + 2a_2 + 6a_3 + 24a_4 + 120a_5$ with $0 \leq a_k \leq k$, so a_5 must equal 5 (in order to obtain 695), and $a_4 \neq 4$ because $5 \cdot 120 + 4 \cdot 24 > 695$. Also a_4 cannot be less than 3, since, for $a_4 = 2$, we have $2 \cdot 24 + 3 \cdot 6 + 2 \cdot 2 + 1 < 95$. $\therefore a_4 = 3$. Check: $5 \cdot 120 + 3 \cdot 24 + 3 \cdot 6 + 2 \cdot 2 + 1 = 695$.

36. (B) From the diagram we have $4a^2 + b^2 = 9$ and $a^2 + 4b^2 = \frac{49}{4}$. $\therefore 5a^2 + 5b^2 = \frac{85}{4}$. $\therefore a^2 + b^2 = \frac{17}{4}$. Since $AB^2 = 4a^2 + 4b^2 = 17$, we have $AB = (17)^{1/2}$.

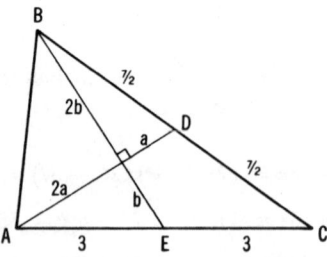

37. (C) Let v_A, v_B, v_C be the respective speeds of A, B, C. Then
(1) $\dfrac{d}{v_A} = \dfrac{d-20}{v_B},$ (2) $\dfrac{d}{v_B} = \dfrac{d-10}{v_C},$ (3) $\dfrac{d}{v_A} = \dfrac{d-28}{v_C},$

SOLUTIONS: 1961 EXAMINATION 57

$$\therefore (4) \quad \frac{d-20}{v_B} = \frac{d-28}{v_C}.$$

Divide equation (4) by d/v_B and use (2) to obtain

$$\frac{d-20}{d} = \frac{d-28}{d-10}, \quad \therefore d = 100.$$

38. (A) Let $AC = h$ and $BC = l$. Then $h^2 + l^2 = 4r^2$. Since $s = h + l$, $s^2 = h^2 + l^2 + 2hl = 4r^2 + 2hl$, where $2hl$ is 4 times the area of $\triangle ABC$. The maximum area of $\triangle ABC$ occurs when $h = l = r\sqrt{2}$.
$\therefore s^2 \leq 4r^2 + 4r^2 = 8r^2$.

39. (B) The greatest distance between two points P_1 and P_2 is the length of a diagonal, $\sqrt{2}$. Place P_1, P_2, P_3, P_4 at the four corners. For the fifth point P_5 to be as far as possible from the other four points, it must be located at the center of the square. The distance from P_5 to any of the other four points is $\frac{1}{2}\sqrt{2}$. Location of the points inside the square will yield distances between pairs of points, the smallest of which is less than $\frac{1}{2}\sqrt{2}$.

40. (A) The minimum value of $(x^2 + y^2)^{1/2}$ is given by the perpendicular OC from O to the line $5x + 12y = 60$. From similar triangles,

$$\frac{OC}{5} = \frac{12}{13}. \quad \therefore OC = \frac{60}{13}.$$

OR

Let $R = (x^2 + y^2)^{1/2}$. Then, since $5x + 12y = 60$, $y = 5 - \frac{5}{12}x$ and

$$R = \left(x^2 + 25 - \frac{25}{6}x + \frac{25}{144}x^2\right)^{1/2}$$

$$= \left[\left(\frac{13}{12}x\right)^2 - \frac{25}{6}x + \left(\frac{25}{13}\right)^2 + 25 - \left(\frac{25}{13}\right)^2\right]^{1/2}$$

$$= \left[\left(\frac{13}{12}x - \frac{25}{13}\right)^2 + 25 - \left(\frac{25}{13}\right)^2\right]^{1/2}.$$

The minimum value of R occurs when $\frac{13}{12}x - \frac{25}{13} = 0$.

$$\therefore R_{\min} = \left[25 - \left(\frac{25}{13}\right)^2\right]^{1/2} = \frac{60}{13}.$$

1962 Solutions

Part 1

1. (D) The numerator is equal to 1, the denominator is $\frac{1}{5} + \frac{1}{3} = \frac{8}{15}$, and $1 \div \frac{8}{15} = \frac{15}{8}$.

2. (A) $\left(\frac{4}{3}\right)^{1/2} - \left(\frac{3}{4}\right)^{1/2} = \frac{2\sqrt{3}}{3} - \frac{\sqrt{3}}{2} = \frac{4\sqrt{3} - 3\sqrt{3}}{6} = \frac{\sqrt{3}}{6}$.

3. (B) Let d be the common difference; then
$$d = (x+1) - (x-1) = 2,$$
and
$$d = (2x+3) - (x+1) = x+2.$$
$$\therefore x + 2 = 2. \quad \therefore x = 0.$$

OR

$$2(x+1) = (x-1) + (2x+3), \quad 2x+2 = 3x+2, \quad x = 0.$$

4. (B) $8^x = 32$. $(2^3)^x = 2^5$, $2^{3x} = 2^5$. $\therefore 3x = 5$, $x = \frac{5}{3}$.

5. (C) The ratio of the circumference of any circle to its diameter is a constant represented by the symbol π.

6. (D) Let s be the length of the side of the triangle; then its area is $(\sqrt{3}/4)s^2 = 9\sqrt{3}$, so $s = 6$. Since its perimeter $3s = 18$ is equal to that of the square, the side of the square is $\frac{18}{4} = \frac{9}{2}$. $\therefore$ the diagonal of the square is $9\sqrt{2}/2$.

7. (C)
$$2\angle b = 180° - \angle B,$$
$$2\angle c = 180° - \angle C,$$
$$\therefore \angle b + \angle c = 180° - \tfrac{1}{2}(\angle B + \angle C).$$

But
$$\angle B + \angle C = 180° - \angle A,$$
$$\therefore \angle b + \angle c = 180° - \tfrac{1}{2}(180° - \angle A),$$

SOLUTIONS: 1962 EXAMINATION

$$\angle BDC = 180° - (\angle b + \angle c)$$
$$= 180° - [180° - \tfrac{1}{2}(180° - \angle A)]$$
$$= \tfrac{1}{2}(180° - \angle A).$$

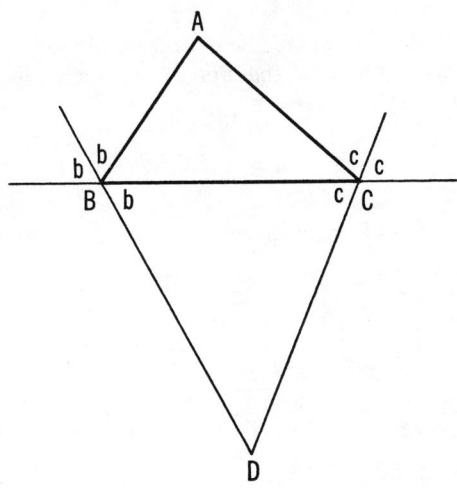

8. **(D)** Let s be the sum and $M(=s/n)$ be the arithmetic mean of the given numbers. Since the number of 1's is $n-1$,

$$s = (n-1)(1) + 1 - \frac{1}{n} = n - \frac{1}{n}.$$

$$\therefore M = \frac{n - 1/n}{n} = 1 - \frac{1}{n^2}.$$

9. **(B)** $x^9 - x = x(x^8 - 1) = x(x^4 + 1)(x^4 - 1)$
$$= x(x^4 + 1)(x^2 + 1)(x + 1)(x - 1).$$

10. **(A)** Let R be the average rate for the trip going. Then

$$R = \frac{150}{3\tfrac{1}{3}} = 45 \text{ (miles per hour)},$$

$$r = \frac{300}{7\tfrac{1}{2}} = 40 \text{ (miles per hour)},$$

$$R - r = 45 - 40 = 5 \text{ (miles per hour)}.$$

11. (B) The roots are
$$\frac{p+1}{2} \quad \text{and} \quad \frac{p-1}{2}.$$

The difference between the larger and the smaller root is 1.

12. (C) The last three terms of the expansion of $(1 - 1/a)^6$ correspond, in inverse order, to the first three terms in the expansion of $(-1/a + 1)^6$. These are

$$1 - 6\cdot\frac{1}{a} + 15\left(-\frac{1}{a}\right)^2,$$

and the sum of the coefficients is $1 - 6 + 15 = 10$.

OR

Expanding completely we have

$$1 - \frac{6}{a} + \frac{15}{a^2} - \frac{20}{a^3} + \frac{15}{a^4} - \frac{6}{a^5} + \frac{1}{a^6};$$

$$15 - 6 + 1 = 10.$$

13. (B) The relation between R, S, and T can be represented by

$$R = k\frac{S}{T} \quad \text{or} \quad \frac{RT}{S} = k,$$

where k is a constant which can be calculated from the given set of values:

$$k = \frac{\frac{4}{3}\cdot\frac{9}{14}}{\frac{3}{7}} = 2.$$

When $R = (48)^{1/2}$, $T = (75)^{1/2}$, then

$$S = \frac{RT}{k} = \frac{(48)^{1/2}(75)^{1/2}}{2} = 30.$$

14. (B) Since

$$1 + a + a^2 + \cdots = \frac{1}{1-a}$$

when $|a| < 1$, the required limiting sum is

$$\frac{4}{1-(-\frac{2}{3})} = 2.4.$$

15. (D) Let the vertex C move along the straight line l_1. For any position C_1 of the vertex on l_1, the centroids (intersection of the three medians) G and G_1 of triangles CAB and C_1AB have the property $CG = \frac{2}{3}CM$ and $C_1G_1 = \frac{2}{3}C_1M$ so the line GG_1 is parallel to line CC_1. (If a line divides two sides of a triangle proportionally, it is parallel to the third side.) Therefore, as C moves along l_1, G moves along l_2 which is a line parallel to l_1.

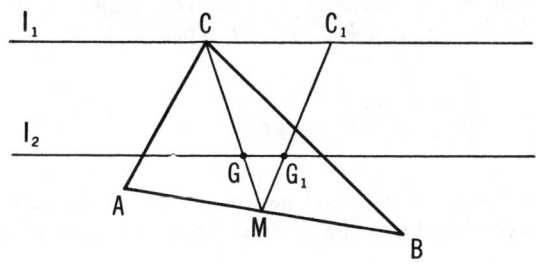

16. (C) The sides of R_1 are 2 inches and 6 inches and the square of its diagonal, d^2, equals $6^2 + 2^2$ ($= 40$). Since the rectangles are similar,

$$\frac{\text{area } R_2}{\text{area } R_1} = \frac{D^2}{d^2}. \quad \therefore \text{ area } R_2 = \frac{225}{40} \cdot 12 = \frac{135}{2}.$$

17. (B) Since $a = \log_8 225$, $\quad 8^a = 225$, $\quad 2^{3a} = 15^2$;

$\qquad\qquad b = \log_2 15$, $\quad 2^b = 15$, $\quad 2^{2b} = 15^2$;

$$\therefore 2^{3a} = 2^{2b}, \quad 3a = 2b, \quad a = \frac{2b}{3}.$$

18. (A) The dodecagon can be decomposed into 12 congruent triangles like triangle OAB shown. (AB is a side of the polygon.)

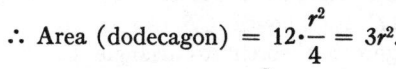

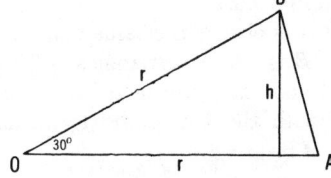

19. (C) By substituting into the given equation the three given sets of x, y-values, we obtain three equations in a, b, and c, namely, $a - b + c = 12$, $c = 5$, and $4a + 2b + c = -3$. Solve the first and third equations for a and b, using for c the value 5 from the second equation. The values obtained are $a = 1$, $b = -6$, $c = 5$, and the sum of these is zero.

20. (A) Let the degree measurement of the angles be represented by $a - 2d$, $a - d$, a, $a + d$, $a + 2d$. Then $5a = 540°$ and $a = 108°$.

Part 2

21. (E) Since the given equation has coefficients which are real numbers, the second root must be $3 - 2i$. Therefore, the product of the roots is

$$\frac{s}{2} = (3 + 2i)(3 - 2i). \quad \therefore s = 26.$$

OR

Substitute $3 + 2i$ for x in the given equation.

$$2(3 + 2i)^2 + r(3 + 2i) + s = 0,$$
$$10 + 3r + s + i(24 + 2r) = 0 + 0 \cdot i.$$
$$\therefore 24 + 2r = 0, \quad r = -12 \quad \text{and} \quad 10 + 3r + s = 0, \quad s = 26.$$

22. (D) $121_b = 1 \cdot b^2 + 2 \cdot b + 1 = b^2 + 2b + 1 = (b + 1)^2.$

The expression $(b + 1)^2$ is, of course, a square number for any value of b. Since, however, the largest possible digit appearing in a number written in the integral base b is $b - 1$, the possible values for b do not include 1 and 2. Therefore, $b > 2$ is the correct answer.

23. (E) When angles A and B are acute and angle C is either acute, obtuse, or right, triangles ABE and CBD are similar. Since AB, CD, and AE are known, CB can be found. Now, by applying the Pythagorean theorem to triangle CBD, where CB and CD are known, we can find DB.

When angle A is obtuse the same analysis holds.

When angle B is obtuse, triangles ABE and ADC are similar. From this fact and the given lengths, CA can be found. Next AD can be found with the aid of the Pythagorean theorem. Finally, $DB = AD - AB$.

When either angle A or angle B is right, the problem is trivial. In the former case, $DB = AB$; in the latter, $DB = 0$.

24. (A) Since $1/x$ represents the fractional part of the job done in 1 hour when the three machines operate together, and so forth,

$$\frac{1}{x+6} + \frac{1}{x+1} + \frac{1}{x+x} = \frac{1}{x}$$

or, more simply,

$$\frac{1}{x+6} + \frac{1}{x+1} - \frac{1}{2x} = 0.$$

$$\therefore 2x(x+1) + 2x(x+6) - (x+6)(x+1) = 0;$$
$$\therefore 3x^2 + 7x - 6 = 0, \quad x = \tfrac{2}{3}.$$

25. (C) $r^2 = 4^2 + (8-r)^2$
    $16r = 80$
    $r = 5.$

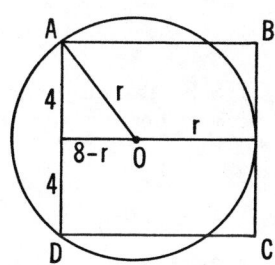

26. (E) Transform the expression $8x - 3x^2$ into

$$3\left(\tfrac{8}{3}x - x^2\right) = 3\left(\tfrac{16}{9} - \tfrac{16}{9} + \tfrac{8}{3}x - x^2\right)$$

$$= 3\left[\tfrac{16}{9} - \left(x - \tfrac{4}{3}\right)^2\right] = \tfrac{16}{3} - 3\left(x - \tfrac{4}{3}\right)^2.$$

The maximum value of this last expression occurs when the term $-3(x - \tfrac{4}{3})^2$ is zero. Therefore, the maximum value of $8x - 3x^2$ is $\tfrac{16}{3}$.

OR

Let $y = -3x^2 + 8x$. The graph of this equation is a parabola with a highest point at $(\tfrac{4}{3}, \tfrac{16}{3})$. Therefore, the maximum value of $-3x^2 + 8x$ is $\tfrac{16}{3}$.

27. (E) The correctness of rule (1) is obvious. To establish rule (2) observe that the left side selects the largest of all three numbers by comparing a with the larger of b and c, while the right side selects the largest of all three by comparing c with the larger of a and b.

To establish rule (3), suppose first that a is the smallest of the three numbers. Then the left side selects it. Each parenthesis on the right selects it and the larger of two equal numbers a is a. If b or c is smallest, say b, then the left side selects the smaller of a and c, and so does the right. If c is the smallest, we argue the same way.

28. (D) Take logarithms of both sides to the base 10.
$$(\log_{10} x)(\log_{10} x) = 3 \log_{10} x - \log_{10} 100$$
or
$$(\log_{10} x)^2 - 3 \log_{10} x + 2 = 0.$$

Solve this quadratic equation in $y = \log_{10} x$. $y^2 - 3y + 2 = 0$, $y = 2$ or 1. $\therefore \log_{10} x = 2$ or 1; $\therefore x = 10^2 = 100$ or $x = 10^1 = 10$.

29. (A) $2x^2 + x < 6$, $x^2 + \dfrac{x}{2} < 3$, $x^2 + \dfrac{x}{2} + \dfrac{1}{16} < 3 + \dfrac{1}{16}$,

$$(x + \tfrac{1}{4})^2 < \tfrac{49}{16}, \qquad |x + \tfrac{1}{4}| < \tfrac{7}{4},$$

that is,
$$x + \tfrac{1}{4} < \tfrac{7}{4} \quad \text{and} \quad x + \tfrac{1}{4} > -\tfrac{7}{4}.$$

$$\therefore x < \tfrac{3}{2} \quad \text{and} \quad x > -2.$$

30. (D) Form I. Since $\sim(p \wedge q) = \sim p \vee \sim q$, statements (2), (3), and (4) are each correct.

Form II. The negation of the statement "p and q are both true" means that either p is true and q is false, or p is false and q is true, or p is false and q is false. Therefore, the statements (2), (3), and (4) are each correct.

SOLUTIONS: 1962 EXAMINATION   65

Part 3

31. **(C)** Let N and n, respectively, represent the numbers of the sides of the two polygons. Then their angles are related by

$$\frac{(N-2)180°}{N} = \frac{3}{2}\frac{(n-2)180°}{n}. \quad \therefore N = \frac{4n}{6-n}.$$

To find N we need try for n only the values 3, 4, 5 (why?). We thus obtain $n = 3$, $N = 4$; $n = 4$, $N = 8$; $n = 5$, $N = 20$, three pairs in all.

32. **(E)** Since $x_{k+1} = x_k + \frac{1}{2}$ and $x_1 = 1$, $x_2 = x_1 + \frac{1}{2} = \frac{3}{2}$,

$$x_3 = x_2 + \frac{1}{2} = \frac{4}{2}, \quad x_4 = x_3 + \frac{1}{2} = \frac{5}{2}, \cdots$$

$$x_n = \frac{n+1}{2}.$$

The required sum is, therefore,

$$\frac{2}{2} + \frac{3}{2} + \frac{4}{2} + \cdots + \frac{n+1}{2}$$

$$= \tfrac{1}{2}(2 + 3 + \cdots + n + n + 1)$$

$$= \tfrac{1}{2}(1 + 2 + \cdots + n) + \frac{n}{2}$$

$$= \frac{1}{2}\frac{n(n+1)}{2} + \frac{n}{2} = \frac{n^2 + 3n}{4}.$$

33. **(A)** The inequality $2 \leq |x - 1| \leq 5$ means that, when $x - 1$ is positive, then $x - 1 \leq 5$ and $x - 1 \geq 2$, and that, when $x - 1$ is negative, then $x - 1 \geq -5$ and $x - 1 \leq -2$. Solve each of these four inequalities.

$$x \leq 6 \quad \text{and} \quad x \geq 3 \quad \text{or} \quad 3 \leq x \leq 6, \text{ and}$$

$$x \geq -4 \quad \text{and} \quad x \leq -1 \quad \text{or} \quad -4 \leq x \leq -1.$$

34. **(E)** Since $x = K^2(x^2 - 3x + 2)$, $K^2x^2 - x(3K^2 + 1) + 2K^2 = 0$.

For x to be a real number the discriminant must be greater than or equal to zero, i.e. $K^4 + 6K^2 + 1 \geq 0$. This is true for all values of K.

35. (B) Let x be the number of degrees the hour hand has moved during the time interval. Then in the same interval, the minute hand has moved $12x$ degrees. But the minute hand has moved $220 + x$ degrees; hence $220 + x = 12x$ or $x = 20$ and the minute hand has moved $220 + x = 240$ degrees. Since a movement of $6°$ corresponds to a time interval of 1 minute, the number of minutes that have elapsed is $240/6 = 40$.

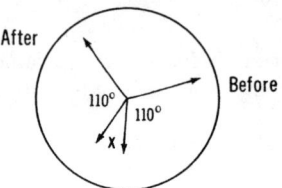

36. (C) $(x - 8)(x - 10) = 2^y$, $x^2 - 18x + 80 - 2^y = 0$.

$$\therefore x = \frac{18 \pm [(18)^2 - 4(80 - 2^y)]^{1/2}}{2} = 9 \pm (1 + 2^y)^{1/2};$$

x is an integer if and only if $1 + 2^y$ is the square of an integer n, that is, $1 + 2^y = n^2$ or

$$2^y = n^2 - 1 = (n - 1)(n + 1).$$

The two factors on the right must be consecutive even integers; for, if they were both odd, their product would be odd and so could not be a power of 2. Moreover, since the product $(n - 1)(n + 1)$ is a power of 2, each factor must be a power of 2. The only pairs of consecutive even integers, each having no prime factors except 2, are $(-4, -2)$ and $(2, 4)$, i.e. $n = \pm 3$. In either case

$$n^2 - 1 = 8 = 2^y, \quad \text{so} \quad y = 3.$$

Consequently, $x = 9 \pm 9^{1/2}$, so $x = 12$ or $x = 6$, and the two solutions (x, y) are $(12, 3)$ and $(6, 3)$.

37. (D) Let x be the common length of AE and AF.

$$\text{Area } (EGC) = \tfrac{1}{2}(1 - x)(1),$$

$$\text{Area } (DFEG) = x\left[\frac{(1 - x) + 1}{2}\right] = \tfrac{1}{2}x(2 - x).$$

$$\therefore \text{Area } (CDFE) = \tfrac{1}{2}(1 + x - x^2)$$

$$= \tfrac{1}{2}[\tfrac{5}{4} - (x^2 - x + \tfrac{1}{4})]$$

$$= \tfrac{5}{8} - \tfrac{1}{2}(x - \tfrac{1}{2})^2.$$

SOLUTIONS: 1962 EXAMINATION   67

The maximum value of this expression is $\frac{5}{8}$. See solution of problem 26.

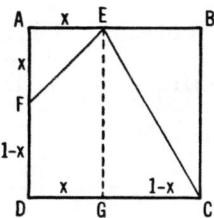

38. **(B)** Let N be the original population. Then
$$N = x^2, \quad N + 100 = y^2 + 1, \quad \text{and} \quad N + 200 = z^2,$$
x, y, z are integers.

Subtract the first equation from the second to obtain
$$100 = y^2 - x^2 + 1 \quad \text{or} \quad y^2 - x^2 = 99.$$
$\therefore (y+x)(y-x) = 99 \cdot 1$ or $33 \cdot 3$ or $11 \cdot 9$. Try $y + x = 99$ and $y - x = 1$; then $y = 50$, $x = 49$. $\therefore N = 49^2 = 2401$, a multiple of 7. Furthermore, $N + 100 = 2501 = 50^2 + 1$ and $N + 200 = 2601 = 51^2$.

With either of the other two sets of factors you obtain a value of N which satisfies the first and second conditions of the problem, but not the third.

39. **(C)** Since the diagonals of the quadrilateral $ADBG$ bisect each other, it is a parallelogram.

$\frac{1}{2}$ Area $(ADBG)$ = Area (ABG) = $\frac{1}{3}$ Area (ABC) = $(15)^{1/2}$,

$\frac{1}{2}$ Area $(ADBG)$ = Area (AGD) = $[s(s-a)(s-g)(s-d)]^{1/2}$,

where a, g, d are the lengths of the sides of $\triangle AGD$, and
$$2s = a + g + d.$$

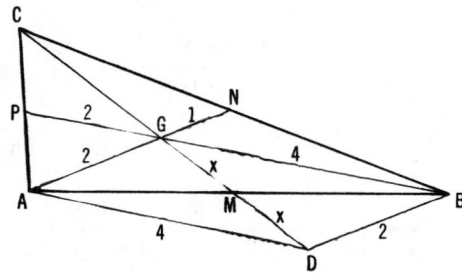

From the given information, we know that
$$a = 2x, \quad g = 4, \quad d = 2, \quad \text{so} \quad s = 3 + x.$$
If we substitute in the above formula, we obtain
$$[(3 + x)(3 - x)(x - 1)(x + 1)]^{1/2} = (15)^{1/2}$$
and, squaring both sides,
$$(x^2 - 1)(9 - x^2) = 15$$
or
$$x^4 - 10x^2 + 24 = (x^2 - 4)(x^2 - 6) = 0;$$
$x^2 = 4$ or 6, so $x = 2$ or $6^{1/2}$ and $CM = 6$ or $3(6)^{1/2}$. However, since the median $BP = 6$, and the triangle has unequal sides, we must reject this solution and retain $CM = 3(6)^{1/2}$.

40. (B) Write the given series as:
$$\frac{1}{10} + \frac{1}{10^2} + \frac{1}{10^3} + \frac{1}{10^4} + \cdots$$
$$+ \frac{1}{10^2} + \frac{1}{10^3} + \frac{1}{10^4} + \cdots$$
$$+ \frac{1}{10^3} + \frac{1}{10^4} + \cdots \text{ and so forth.}$$

Let s_1 be the limiting sum of the first-row series, s_2, that of the second-row series, and so forth.
$$s_1 = \frac{1/10}{1 - (1/10)} = \frac{1}{9},$$
$$s_2 = \frac{(1/10^2)}{1 - (1/10)} = \frac{1}{90},$$
$$s_3 = \frac{(1/10^3)}{1 - (1/10)} = \frac{1}{900},$$
and so forth. Therefore, the required limiting sum equals
$$\frac{1}{9} + \frac{1}{90} + \frac{1}{900} + \cdots = \frac{1/9}{1 - (1/10)} = \frac{10}{81}.$$

1963 Solutions

Part 1

1. (D) Since
$$\frac{x}{x+1}$$
is not defined when the denominator is zero, and since the denominator $x + 1$ equals zero when $x = -1$, any point whose abscissa is -1, such as $(-1, 1)$, can not be on the graph.

2. (A) $n = 2 - (-2)^{2-(-2)} = 2 - (-2)^4 = 2 - 16 = -14$.

3. (E) Since
$$\frac{1}{x+1} = x - 1, \qquad x^2 - 1 = 1, \qquad x^2 = 2, \qquad x = +\sqrt{2} \text{ or } -\sqrt{2}.$$

4. (D) For $x^2 = 3x + k$ to have two equal roots, its discriminant $9 + 4k$ must equal zero; $\therefore k = -\frac{9}{4}$.

5. (E) Choices (A), (B), and (D) must be rejected since the logarithm of a negative real number is not defined (on this level of mathematical study). Choice (C) must be rejected since $\log 1 = 0$. Choice (E) is the correct one since $\log x$, for $0 < x < 1$, exists and is a negative real number.

   OR

   Since it is given that $\log_{10} x < 0$, then $x < 10^0 \, (=1)$. This result, together with the fact that (on this level) $\log x$ is defined only for $x > 0$, leads to choice (E).

6. (B) BC is the median to the hypotenuse AD, and is, therefore, equal to one-half the hypotenuse. $\therefore AB = BC = AC$, and, consequently, the magnitude of angle DAB is $60°$.

7. (A) Two lines are perpendicular when the product of their slopes is -1. Since the slopes are, respectively, $\frac{2}{3}$, $-\frac{2}{3}$, $-\frac{2}{3}$, $-\frac{3}{2}$, and

since $(\frac{2}{3})(-\frac{3}{2}) = -1$, lines (1) and (4) are perpendicular.

OR

Two lines, $a_1x + b_1y = c_1$ and $a_2x + b_2y = c_2$, are perpendicular if $a_1a_2 + b_1b_2 = 0$. In (1) $a_1 = -2$, $b_1 = 3$, and in (4) $a_2 = 3$, $b_2 = 2$, so that $a_1a_2 + b_1b_2 = -6 + 6 = 0$.

8. (D) $1260x = (2 \cdot 2 \cdot 3 \cdot 3 \cdot 5 \cdot 7)x$, and this is a cube only if the exponent of each of its different prime factors is 3. Therefore,

$$x = 2 \cdot 3 \cdot 5^2 \cdot 7^2 = 7350$$

is the smallest x such that $1260x$ is a cube.

9. (C) $\left(a - \dfrac{1}{a^{1/2}}\right)^7 = a^7 - 7a^{11/2} + 21a^4 - 35a^{5/2}$

$$+ 35a - 21a^{-1/2} + 7a^{-2} - a^{-7/2}.$$

OR

$$(r+1)\text{th term} = \binom{7}{r}a^{7-r}(-a^{-1/2})^r = \pm\binom{7}{r}a^{7-r-r/2}.$$

We need

$$a^{7-r-r/2} = a^{-1/2}, \qquad 7 - \frac{3r}{2} = -\frac{1}{2}, \qquad r = 5.$$

$$\therefore \text{6th term} = \frac{7 \cdot 6 \cdot 5 \cdot 4 \cdot 3}{1 \cdot 2 \cdot 3 \cdot 4 \cdot 5} a^2 (-a)^{-5/2} = -21a^{-1/2}$$

10. (B) $(a/2)^2 + (a-d)^2 = d^2$, $\quad d = 5a/8$.

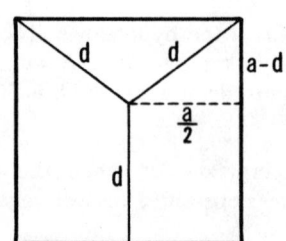

SOLUTIONS: 1963 EXAMINATION   71

11. (B) Since the arithmetic mean of the 50 numbers is 38, the sum of the 50 numbers is $50 \times 38 = 1900$. When the two numbers 45 and 55 are removed, 48 numbers with a sum of 1800 remain. Therefore, their arithmetic mean is $\frac{1800}{48} = 37.5$.

OR

$a_1 + a_2 + \cdots + a_{48} + 100 = 50 \cdot 38 = 50(36 + 2) = 50 \cdot 36 + 50 \cdot 2.$

$\therefore a_1 + a_2 + \cdots + a_{48} = 50 \cdot 36 = 48 \cdot (\text{new A.M.}).$

$$\therefore (\text{new}) \text{ A.M.} = \frac{50 \cdot 36}{48} = 37.5.$$

12. (E) The sum of the x-coordinates of P and R equals the sum of the x-coordinates of Q and S, and, similarly, for the y-coordinates.
$\therefore x + 1 = 6$, $x = 5$ and $y - 5 = -1$, $y = 4$. $\therefore x + y = 9$.

OR

Let S_x be the x-coordinate of S and S_y, the y-coordinate of S, and, similarly, for P, Q, and R. Then, by congruent triangles, $Q_x - R_x = P_x - S_x$, $1 - 9 = -3 - S_x$, $S_x = 5$. Similarly $S_y = 4$. $\therefore S_x + S_y = 9$.

13. (E) We shall show successively that it is impossible that (i) precisely one of the integers a, b, c, d be negative, (ii) that precisely two be negative, (iii) that precisely three be negative, (iv) that all four be negative. We may then conclude that none of the four integers can be negative. Whenever we assume that a, b, c or d is negative, we shall set $-a = \alpha$, $-b = \beta$, $-c = \gamma$, $-d = \delta$ so that the greek letters will always denote positive integers.

(i) If exactly one of the numbers were negative, one side of the given equation $2^a + 2^b = 3^c + 3^d$ would be an integer, the other not; so this possibility is ruled out.

(ii)(a) Suppose $a < 0$, $b < 0$ and $c \geq 0$, $d \geq 0$; then the right side is an integer ≥ 2 while the biggest value the left side can have is 1 (for $a = b = -1$).

(b) If $c < 0$, $d < 0$ and $a \geq 0$, $b \geq 0$, the left side is at least 2 while the right is at most $\frac{2}{3}$.

(c) If $a < 0$, $c < 0$ and $b \geq 0$, $d \geq 0$, we may write

$$\frac{1}{2^\alpha} - \frac{1}{3^\gamma} = \frac{3^\gamma - 2^\alpha}{2^\alpha 3^\gamma} = 3^d - 2^b = n \quad (\text{an integer})$$

or

$$3^\gamma - 2^\alpha = 2^\alpha 3^\gamma n.$$

The right side of this equation is even, the left is odd (since it is the difference of an odd and an even number. The case $b < 0$, $d < 0$, $a \geq 0$, $c \geq 0$ may be handled in exactly the same manner.

(iii) It suffices to examine the two cases

(A) only $a \geq 0$, (B) only $c \geq 0$.

In case A, we write

$$2^a = \frac{1}{3^\gamma} + \frac{1}{3^\delta} - \frac{1}{2^\beta} = \frac{2^\beta 3^\delta + 2^\beta 3^\gamma - 3^{\gamma+\delta}}{2^\beta 3^{\gamma+\delta}}$$

$$= n \quad \text{(an integer)}$$

or

$$2^\beta(3^\delta + 3^\gamma) - 3^{\gamma+\delta} = 2^\beta 3^{\gamma+\delta} n;$$

the integer on the left is odd, that on the right is even. In case B, we write

$$3^c = \frac{1}{2^\alpha} + \frac{1}{2^\beta} - \frac{1}{3^\delta} = \frac{2^\beta 3^\delta + 2^\alpha 3^\delta - 2^{\alpha+\beta}}{2^{\alpha+\beta} 3^\delta}$$

$$= n \quad \text{(an integer)}$$

or

$$3^\delta(2^\alpha + 2^\beta) - 2^{\alpha+\beta} = 2^{\alpha+\beta} 3^\delta n;$$

the integer on the right is divisible by 3, that on the left is not.

(iv) If all integers are negative, we write

$$\frac{1}{2^\alpha} + \frac{1}{2^\beta} = \frac{1}{3^\gamma} + \frac{1}{3^\delta} \quad \text{or} \quad (2^\alpha + 2^\beta)3^{\gamma+\delta} = (3^\gamma + 3^\delta)2^{\alpha+\beta}.$$

If $\alpha \neq \beta$, say $\alpha < \beta$, reduce the equation to

$$(1 + 2^{\beta-\alpha})3^{\gamma+\delta} = (3^\gamma + 3^\delta)2^\beta,$$

where the left side is odd, the right even.
If $\alpha = \beta$, our equation becomes

$$2^{\alpha+1}3^{\gamma+\delta} = (3^\gamma + 3^\delta)2^{2\alpha};$$

when $\alpha = 1$, this is equivalent to $3^\gamma \cdot 3^\delta = 3^\gamma + 3^\delta$ which is false for all γ, δ. When $\alpha > 1$, we have $\alpha - 1 > 0$, so our equation implies

$$3^{\gamma+\delta} = 2^{\alpha-1}(3^\gamma + 3^\delta),$$

which states that an odd number is equal to an even number.

All possibilities have now been ruled out.

SOLUTIONS: 1963 EXAMINATION

14. (A) Let the roots of the first equation be r and s. Then $r + s = -k$ and $rs = 6$. Then, for the second equation, $r + 5 + s + 5 = k$ and $(r + 5)(s + 5) = 6$. Therefore $rs + 5(r + s) + 25 = 6$, $6 + 5(-k) + 25 = 6$, $k = 5$.

    OR

    Let r be one of the roots of the first equation; then $r + 5$ is the associated root of the second equation.

    $$\therefore (r + 5)^2 - k(r + 5) + 6 = 0,$$

    $$r^2 + (10 - k)r + 31 - 5k = 0.$$

    But r satisfies the first equation so that $r^2 + kr + 6 = 0$.

    $$\therefore 10 - k = k \quad \text{and} \quad 31 - 5k = 6.$$

    Each of these equations leads to the result $k = 5$.

15. (C) Let S, s, r, respectively, represent the magnitudes of a side of the equilateral triangle, a side of the square, and a radius of the circle. Since $S = 2r\sqrt{3}$ and $s = r\sqrt{2}$, the area of the triangle is

    $$\frac{(2r\sqrt{3})^2\sqrt{3}}{4} \quad (= 3\sqrt{3}r^2)$$

    and the area of the square is $(r\sqrt{2})^2$ $(= 2r^2)$. The required ratio is, therefore, $3\sqrt{3}:2$.

16. (C) Since a, b, c are in A.P., $2b = a + c$. Since $a + 1, b, c$ are in G.P., $b^2 = c(a + 1)$. Similarly, $b^2 = a(c + 2)$. $\therefore c = 2a$, $a = \frac{2}{3}b$, $b^2 = \frac{8}{9}b^2 + \frac{4}{3}b$. $\therefore b = 12$.

17. (A) The expression is not defined for $y = a$ and for $y = -a$. For all other values, we may simplify the fraction by multiplying numerator and denominator by $(a + y)(a - y)$, obtaining

    $$\frac{a^2 - ay + ay + y^2}{ay - y^2 - a^2 - ay} = \frac{a^2 + y^2}{-(a^2 + y^2)} = -1.$$

18. (A) Triangle EFA is a right triangle (since EF is a diameter) with angle E as one of its acute angles. Triangle EUM is a right triangle with angle E as one of its acute angles. Therefore, triangle $EFA \sim$ triangle EUM.

19. (B) $\dfrac{49 + \frac{7}{8}(n-50)}{n} \geq \dfrac{9}{10}.$ $\therefore n \leq 210.$

20. (D) Let $h =$ number of hours in which they meet. Then
$$\tfrac{9}{2}h + \tfrac{1}{2}h[2\cdot\tfrac{13}{4} + (h-1)\tfrac{1}{2}] = 76, \quad h^2 + 30h - 304 = 0, \quad h = 8.$$
The meeting point is, therefore, 36 miles from R and 40 miles from S, so that $x = 4$. How would you solve this problem without the assumption that h is an integer?

Part 2

21. (E) $x^2 - y^2 - z^2 + 2yz + x + y - z$
$$= x^2 - (y^2 - 2yz + z^2) + x + y - z$$
$$= x^2 - (y-z)^2 + x + y - z$$
$$= (x + y - z)(x - y + z) + x + y - z$$
$$= (x + y - z)(x - y + z + 1).$$

22. (D)   Minor arc $AC = 360° - (120° + 72°) = 168°$,
and
$$\angle BOE = 72° + 168°/2 = 156°.$$
In isosceles triangle EOB, each base angle is $\tfrac{1}{2}(180° - 156°) = 12°$. $\angle BAC = \tfrac{1}{2}\cdot 72° = 36°$, so
$$\dfrac{\angle OBE}{\angle BAC} = \dfrac{12}{36} = \dfrac{1}{3}.$$

23. (B) If a, b, c, respectively, represent the initial amounts of A, B, C, then the given conditions lead to the following:

After Transaction	A has	B has	C has
I	$a - b - c$	$2b$	$2c$
II	$2(a-b-c)$	$2b - (a-b-c) - 2c$ $(= 3b - a - c)$	$4c$
III	$4(a-b-c)$	$2(3b-a-c)$	$4c - 2(a-b-c) - (3b-a-c)$ $(= 7c - a - b)$

SOLUTIONS: 1963 EXAMINATION 75

Consequently

$$4(a - b - c) = 16, \quad 6b - 2a - 2c = 16, \quad 7c - a - b = 16.$$

Solving this set of equations, you obtain $a = 26$, $b = 14$, $c = 8$.

OR

Working from the last condition to the first, we may set up the following table:

Amounts	Step 4	Step 3	Step 2	Step 1
a	16	8	4	26 (required)
b	16	8	28	14
c	16	32	16	8

24. (B) If $b^2 - 4c \geq 0$, then the roots are real. Tabulating the possibilities, we have 19 in all as follows:

When c is selected to be:	6	6	5	5	4	4	4	3	3	3	2	2	2	2	1	1	1	1	1
The possible values for b are:	5	6	5	6	4	5	6	4	5	6	3	4	5	6	2	3	4	5	6
The number of possible values for b are:	2		2		3			3			4				5				

25. (A) $\triangle CDF \cong \triangle CBE$ ($CD = CB$, angle DCF = angle BCE),

$$\therefore CF = CE.$$

$$\text{Area } (\triangle CEF) = \tfrac{1}{2} CE \cdot CF = \tfrac{1}{2} CE^2 = 200.$$

$$\therefore CE^2 = 400.$$

$$\text{Area (square)} = CB^2 = 256.$$

Since $BE^2 = CE^2 - CB^2 = 400 - 256 = 144$, $BE = 12$.

26. (E) The implication $(p \to q) \to r$ is true when (i) the consequent, r, is true and the antecedent, $p \to q$, is either true or false, and (ii) the consequent is false and the antecedent is false.

The implication $p \to q$ is true when (i) the consequent, q, is true and the antecedent, p, is either true or false, and (ii) the consequent is false and the antecedent is false.

In (1) $p \to q$ is false and r is true so that $(p \to q) \to r$ is true.
In (2) $p \to q$ is true and r is true, so that $(p \to q) \to r$ is true.
In (3) $p \to q$ is false and r is false so that $(p \to q) \to r$ is true.
In (4) $p \to q$ is true and r is true, so that $(p \to q) \to r$ is true.

27. (C) Let n be the number of lines and r the number of regions. It is not too difficult to discover the rule that $r = \frac{1}{2}n(n+1) + 1$. For $n = 6$, $r = \frac{1}{2} \cdot 6 \cdot 7 + 1 = 22$. Hint: (1) for one line there are two regions, that is, one more than the number of lines; (2) note what happens to the regions when a line is added.

28. (D) The product of two numbers whose sum is a fixed quantity is maximized when each of the numbers is one-half the sum. Since the sum of the roots is $\frac{4}{3}$, the maximum product, $\frac{4}{9}$, is obtained when each of the roots is $\frac{2}{3}$. Therefore, $k/3 = \frac{4}{9}$, so that $k = \frac{4}{3}$.

OR

Solving the given equation for $k/3$ (the product of the roots), we have $k/3 = \frac{4}{3}x - x^2 = \frac{4}{9} - (\frac{2}{3} - x)^2$. The right side of this equation is a maximum when $x = \frac{2}{3}$, so that $k/3$ is a maximum when $x = \frac{2}{3}$. $\therefore k/3$ (max) $= \frac{4}{3} \cdot \frac{2}{3} - \frac{4}{9} = \frac{4}{9}$, $\therefore k$ (max) $= \frac{4}{3}$.

OR

The product of the roots is $k/3$. We seek the largest possible k consistent with real roots of the given equation. For real roots $16 - 12k \geq 0$, so that $\frac{4}{3} \geq k$. Hence the desired k is $\frac{4}{3}$.

29. (C) The abscissa of the maximum (or minimum) point on the parabola $ax^2 + bx + c$ is $-b/2a$. For the given parabola $160t - 16t^2$, $-b/2a = -160/-32 = 5$, and the value of s when $t = 5$ is 400.

OR

Since
$$s = 160t - 16t^2,$$
$$s = 400 - 400 + 160t - 16t^2 = 400 - 16(5 - t)^2.$$

The right side of this equation is a maximum when $5 - t = 0$. Therefore, $s(\max) = 400$.

30. (C) Since
$$\frac{1 + \dfrac{3x + x^3}{1 + 3x^2}}{1 - \dfrac{3x + x^3}{1 + 3x^2}} = \frac{1 + 3x + 3x^2 + x^3}{1 - 3x + 3x^2 - x^3} = \frac{(1+x)^3}{(1-x)^3} = \left(\frac{1+x}{1-x}\right)^3,$$

we have
$$G = \log\left(\frac{1+x}{1-x}\right)^3 = 3\log\frac{1+x}{1-x} = 3F.$$

Part 3

31. (D) Solving $2x + 3y = 763$ for x, we have
$$x = \frac{763 - 3y}{2}.$$

Since x is a positive integer, $763 - 3y$ must be a positive even number, so that y must be a positive odd integer, such that $3y \leq 763$. There are 254 multiples of 3 less than 763, half of which are even multiples and half, odd multiples. Therefore, there are 127 possible solutions to the given equation under the stated conditions.

OR

$2x + 3y = 763, \quad x + y + \tfrac{1}{2}y = 381 + \tfrac{1}{2}, \quad \therefore \tfrac{1}{2}y = N + \tfrac{1}{2},$

$y = 2N + 1; \quad \therefore x = 380 - 3N.$

For $380 - 3N > 0$, $N < 126\tfrac{2}{3}$ so that $N = 0, 1, 2, \cdots, 126$, a total of 127 solutions.

32. (A) From the given conditions we have

(1) $3(x + y) = a + b$ and (2) $3xy = ab$.

Divide the left side of equation (1) by $3xy$ and the right side by ab:
$$\frac{1}{y} + \frac{1}{x} = \frac{1}{b} + \frac{1}{a}.$$

This is impossible since $x < a < b$ and $y < a < b$.

33. (A) There are two possibilities, L_1 and L_2. The methods for finding the equation of L_2 are similar to the methods for finding the equation of L_1 shown here: Since L_1 is parallel to the line $y = \tfrac{3}{4}x + 6$, its equation is $y = \tfrac{3}{4}x + b$, where b, the y-intercept, is to be determined. Since $\triangle ABC \sim \triangle DAO$ (see figure on page 78),
$$\frac{AB}{AD} = \frac{AC}{DO}, \quad \text{so that} \quad \frac{AB}{10} = \frac{4}{8} \quad \text{and} \quad AB = 5.$$

Hence $b = 1$, and the equation of L_1 is $y = \tfrac{3}{4}x + 1$.

OR

Let d_1 be the distance from the origin to L_1 and let d be the distance from the origin to the given line. Then

$$d_1 = \frac{4b}{5} \quad \text{and} \quad d = \frac{24}{5}. \quad \therefore d - d_1 = \frac{24}{5} - \frac{4b}{5} = 4, \quad \text{and} \quad b = 1.$$

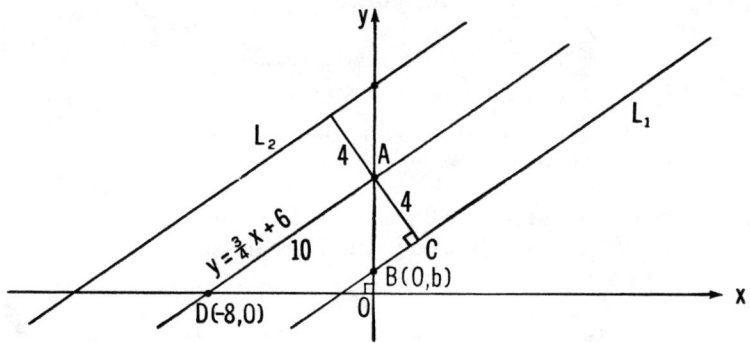

34. (B) The vertex angle C opposite the base c of isosceles triangle ABC with $a = b = \sqrt{3}$ increases as c increases. Since $c > 3$, C is greater than the vertex angle C' in triangle $A'B'C'$ with $a' = b' = \sqrt{3}$ and $c' = 3$. The altitude $C'D$ has length

$$[3 - (\tfrac{3}{2})^2]^{1/2} = \frac{\sqrt{3}}{2} = \frac{a'}{2},$$

so $\angle A = \angle B = 30°$, and $\angle C = 120° = x$.

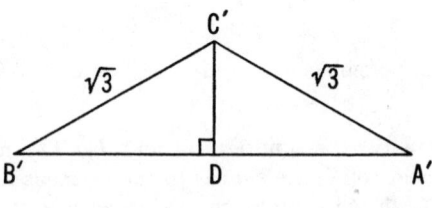

OR

The altitude CD of triangle ABC bisects $\angle C$; since $c > 3$,

$$\sin \tfrac{1}{2}C = \frac{c/2}{\sqrt{3}} \geq \frac{\sqrt{3}}{2}.$$

SOLUTIONS: 1963 EXAMINATION   79

Angle $\tfrac{1}{2}C$ is acute since $\angle C < 180°$, and the sine of an acute angle is an increasing function. Hence

$$\tfrac{1}{2}C > \arc\sin \frac{\sqrt{3}}{2} = 60°,$$

so $x = 120°$.

OR

$c^2 = a^2 + b^2 - 2ab \cos C,$

$$\cos C = \frac{a^2 + b^2 - c^2}{2ab} = \frac{3 + 3 - c^2}{2\sqrt{3}\sqrt{3}} = \frac{6 - c^2}{6}$$

where $c^2 > 9$. $\therefore \cos C < -\tfrac{1}{2}$, so that $\angle C > 120°$.

35. (B) Let the lengths of the sides be 21, x, $27 - x$, and let the altitude to the given side divide it into segments of lengths y and $21 - y$. Then the square of the altitude may be expressed by

$$x^2 - y^2 = (27 - x)^2 - (21 - y)^2.$$

Simplifying, we obtain $48 - 9x + 7y = 0$.

$$\therefore x = 5 + \frac{3 + 7y}{9}.$$

The smallest integer y that makes x an integer is $y = 6$, and, for this value, $x = 10$. For the next admissible value $y = 15$, $x = 17$, so that $27 - x = 10$. For larger values of y, we obtain sides which cannot belong to triangles.

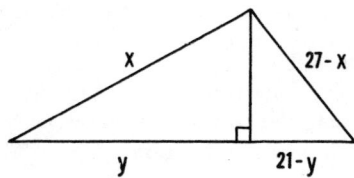

36. (C) Let M_0 represent the number of cents at the start. Suppose the first bet results in a win; then the amount at the end of the first bet is $\tfrac{1}{2}M_0 + M_0 = \tfrac{3}{2}M_0$. Let $M_1 = \tfrac{3}{2}M_0$. Suppose the second bet results in a win; then the amount at the end of the second bet is $\tfrac{1}{2}M_1 + M_1 = \tfrac{3}{2}M_1 = (\tfrac{3}{2})^2 M_0$. Hence, a win at any stage results in an amount $\tfrac{3}{2}$ times the amount at hand at that stage.

If the first bet results in a loss, the amount at the end of the first bet is $\tfrac{1}{2}M_0$. Let $m = \tfrac{1}{2}M_0$. Suppose the second bet results in a loss, then the amount at the end of the second bet is $m/2 = (\tfrac{1}{2})^2 M_0$. Hence, a loss at any stage results in an amount $\tfrac{1}{2}$ times the amount at hand at that stage.

For three wins and three losses, in any order, the amount left is $(\tfrac{3}{2})^3(\tfrac{1}{2})^3 M_0 = \tfrac{27}{64} M_0$. Consequently, the amount lost is $M_0 - \tfrac{27}{64} M_0 = \tfrac{37}{64} M_0$. Since $M_0 = 64$ cents, the amount lost is 37 cents.

37. (D) Consider first a single segment $P_1 P_2$; in this case the point P can be selected anywhere in the segment since, for any such selection, $s = PP_1 + PP_2 = P_1 P_2$. Now consider the case of two segments with end-points P_1, P_2, P_3; the smallest value of s is obtained by choosing P to coincide with P_2 since, for this selection,

$$s = P_1 P_2 + P_2 P_3$$

whereas, for any other selection, $s > P_1 P_2 + P_2 P_3$. These two cases are quite typical. In fact, for any even number $n = 2k$ of points, P should be located somewhere on the segment $P_k P_{k+1}$; and for an odd number $n = 2k + 1$ of points, P should coincide with P_{k+1}. To see this, compute the sum of distances PP_i when P is in the position suggested above and observe that, when P is moved either to the right or to the left, that sum increases.

OR

We now give a lengthy but systematic solution. Suppose the points P_i ($i = 1, 2, \cdots, 7$) lie along the x-axis and have coordinates x_i. Let x be the coordinate of an arbitrary point P; then the distance from P to P_k is $|x - x_k|$. We are to determine the position of P so that

$$|x - x_1| + |x - x_2| + \cdots + |x - x_7|$$

is as small as possible.

We shall solve this problem for any number n of points and specialize the solution to the case $n = 7$. We first find the value of x which minimizes the function

$$f(x) = |x - x_1| + |x - x_2| + \cdots + |x - x_n| = \sum_{i=1}^{n} |x - x_i|$$

for given numbers x_i such that $x_1 < x_2 < \cdots < x_n$. We claim that $f(x)$ is a continuous piecewise linear function; that is, its graph is made up of line segments. To see this, recall that

$$|a - b| = \begin{cases} a - b & \text{if } a \geq b \\ b - a & \text{if } a < b \end{cases}$$

and write $f(x)$ as follows:

for $x < x_1$,
$$f(x) = x_1 - x + x_2 - x + \cdots + x_n - x$$
$$= -nx + \sum_{i=1}^{n} x_i$$

for $x_1 \leq x < x_2$,
$$f(x) = x - x_1 + x_2 - x + \cdots + x_n - x$$
$$= (2-n)x + \sum_{i=1}^{n} x_i - 2x_1$$

....

for $x_k \leq x < x_{k+1}$,
$$f(x) = x - x_1 + \cdots + x - x_k + x_{k+1} - x + \cdots + x_n - x$$
$$= (2k - n)x + \sum_{i=1}^{n} x_i - 2\sum_{i=1}^{k} x_i$$

....

for $x_n < x$,
$$f(x) = x - x_1 + x - x_2 + \cdots + x - x_n = nx - \sum_{i=1}^{n} x_i.$$

In each interval, $f(x)$ is of the form $mx + b$, hence a piece of a line with slope m. As x goes from left to right, the slopes take the integer values $-n$, $2 - n$, $\cdots$, $2k - n$, $\cdots$, n, that is, as x passes one of the values x_k, the slope of the graph jumps by $+2$. When the slope is negative, $f(x)$ decreases; when it is positive, $f(x)$ increases. To see that f is continuous, either verify from the above formulas that the line segments meet at endpoints x_k of intervals, or simply notice that each distance $|x - x_j|$ is a continuous function and so the sum of such distances is continuous. $f(x)$ has a minimum when it stops decreasing and begins increasing.

If n is even, say $n = 2m$, then the graph of $f(x)$ has a horizontal segment from x_m to x_{m+1} (the slope is $2m - 2m = 0$) along which $f(x)$ is least. If n is odd, say $n = 2m + 1$, $f(x)$ has a lowest value at x_{m+1}, where the graph has a corner; x_{m+1} separates the decreasing part (ending with slope $2m - (2m+1) = -1$) from the increasing part (beginning with slope $2(m+1) - (2m+1) = 1$).

We conclude: If n is even, the point P that minimizes f may be taken anywhere on the segment $P_{n/2}P_{(n/2)+1}$, and if n is odd, P must be taken at $P_{(n+1)/2}$. In particular, if $n = 7$, choose P to coincide with P_4.

(The accompanying figure represents the graph of $f(x)$ for a special case of 5 points.)

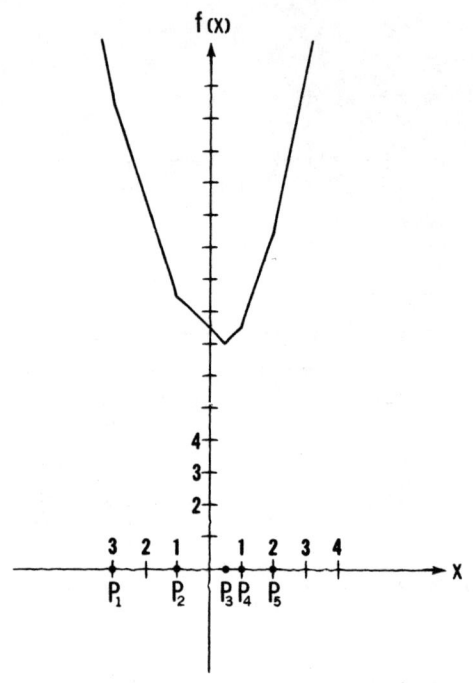

38. (E) Let $BE = x$, $DG = y$, $AB = b$. Since $\triangle BEA \sim \triangle GEC$,

$$\frac{8}{x} = \frac{b-y}{b}, \quad b - y = \frac{8b}{x}, \quad y = b - \frac{8b}{x} = \frac{b(x-8)}{x}.$$

Since $\triangle FDG \sim \triangle BCG$,

$$\frac{24}{x+8} = \frac{y}{b-y}, \quad \frac{24}{x+8} = \frac{b(x-8)}{x \cdot (8b/x)} = \frac{x-8}{8},$$

$$x^2 - 64 = 192, \quad x = 16.$$

Note: Try to prove generally that

$$\frac{1}{BE} = \frac{1}{BG} + \frac{1}{BF}.$$

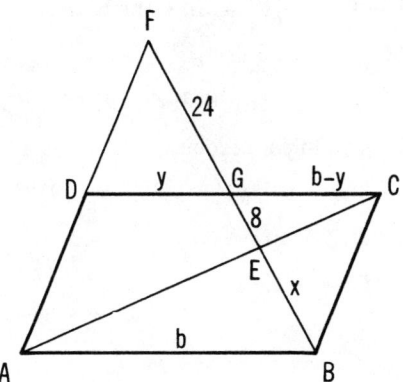

39. (D) Draw $DR \parallel AB$.

$$\frac{CR}{RE} = \frac{CD}{DB} = \frac{3}{1}, \quad \frac{RD}{EB} = \frac{CD}{CB} = \frac{3}{4};$$

$$\therefore CR = 3RE = 3RP + 3PE \text{ and } RD = \tfrac{3}{4}EB,$$

$$\therefore CP = CR + RP = 4RP + 3PE.$$

Since $\triangle RDP \sim \triangle EAP$,

$$\frac{RP}{PE} = \frac{RD}{AE}, \quad \therefore RD = \frac{RP \times AE}{PE}.$$

But $AE = \tfrac{3}{2}EB$.

$$\therefore RD = \frac{RP}{PE} \cdot \tfrac{3}{2}EB.$$

$$\therefore \tfrac{3}{4}EB = \tfrac{3}{2}EB \cdot \frac{RP}{PE}, \quad RP = \tfrac{1}{2}PE,$$

$$CP = 4 \cdot \tfrac{1}{2}PE + 3PE = 5PE; \quad \therefore \frac{CP}{PE} = 5.$$

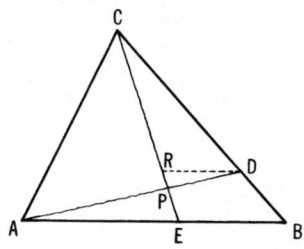

40. (C) $(x+9)^{1/3} - (x-9)^{1/3} = 3$. Cube both sides and obtain
$$x + 9 - 3(x+9)^{2/3}(x-9)^{1/3}$$
$$+ 3(x+9)^{1/3}(x-9)^{2/3} - x + 9 = 27$$
which, when simplified, becomes
$$9 = -3(x+9)^{1/3}(x-9)^{1/3}[(x+9)^{1/3} - (x-9)^{1/3}]$$
$$= -3(x^2 - 81)^{1/3}[3].$$
$\therefore (x^2 - 81)^{1/3} = -1$, $x^2 = 80$.

OR

Let $u = (x+9)^{1/3}$, $v = (x-9)^{1/3}$. then $u - v = 3$,
$(u-v)^2 = u^2 - 2uv + v^2 = 9$, and
$$u^3 - v^3 = x + 9 - (x-9) = 18 = (u-v)(u^2 + uv + v^2)$$
$$= 3(u^2 + uv + v^2),$$
$$\therefore u^2 + uv + v^2 = 6.$$
But $u^2 - 2uv + v^2 = 9$,
$$\therefore 3uv = -3, \quad uv = -1. \quad \text{Also,}$$
$$(u+v)^2 = u^2 + 2uv + v^2 = (u-v)^2 + 4uv = 9 - 4 = 5,$$
$u + v = 5^{1/2}$, $\quad u - v = 3$, $\quad$ so $\quad 2u = 3 + 5^{1/2}$.
Hence
$$(2u)^3 = 8u^3 = 72 + 32(5)^{1/2} = 8(x+9) = 8x + 72$$
$\therefore x = 4(5)^{1/2}$, $\quad x^2 = 80$.

1964 Solutions

Part 1

1. (E) $[\log_{10} (5 \log_{10} 100)]^2 = [\log_{10} (5 \cdot 2)]^2 = 1^2 = 1.$

2. (C) $\qquad x^2 - 4y^2 = (x + 2y)(x - 2y) = 0.$
   $\therefore x + 2y = 0 \quad \text{or} \quad x - 2y = 0.$

   The graph of each equation is a straight line, so that the correct answer is (C).

3. (D) Dividend = Divisor·Quotient + Remainder. From $x = uy + v$, it follows that $x + 2uy = 3uy + v$, so v is the remainder.

   OR

   Since the remainder when x is divided by y is v, and the remainder when $2uy$ is divided by y is zero, the remainder when $x + 2uy$ is divided by y is the same as the remainder when x is divided by y, to wit, v.

4. (A) Since $P = x + y$ and $Q = x - y$, $P + Q = 2x$, $P - Q = 2y$.
   $$\therefore \frac{P+Q}{P-Q} - \frac{P-Q}{P+Q} = \frac{2x}{2y} - \frac{2y}{2x} = \frac{x^2 - y^2}{xy}.$$

5. (A) $y = kx$, k is a constant determined from the given data: $8 = k \cdot 4$, $k = 2$; when $x = -8$, $y = 2 \cdot (-8) = -16$.

6. (B) The common ratio is
   $$\frac{2x+2}{x} = \frac{3x+3}{2x+2} = \frac{3}{2}, \qquad x \neq 0,\ x \neq -1.$$
   $\therefore 2x + 2 = \frac{3}{2}x$, $x = -4$. $\therefore$ the 4th term is $(-4)(\frac{3}{2})^3 = -13\frac{1}{2}.$

7. (C) For equal roots the discriminant, $p^2 - 4p$, is zero. This condition is satisfied by $p = 0$ and $p = 4$, so that (C) is the correct answer.

8. **(C)** We can re-write the equation as

$$(x - \tfrac{3}{4})[(x - \tfrac{3}{4}) + (x - \tfrac{1}{2})] = (x - \tfrac{3}{4})(2x - \tfrac{5}{4}) = 0.$$

$x = \tfrac{3}{4}$ or $x = \tfrac{5}{8}$. Since $\tfrac{5}{8} < \tfrac{3}{4}$, the correct answer is (C).

9. **(E)** The cost of the article is $\tfrac{7}{8} \times \$24 = \$21$. A gain of $33\tfrac{1}{3}\%$ of the cost is $\tfrac{1}{3} \times \$21 = \$7$. The article must, therefore, be sold for $21 + 7 = 28$ (dollars). Let M be the marked price, in dollars; $M - \tfrac{1}{5}M = 28$. $\therefore M = 35$, so that the correct answer is (E).

10. **(A)** For the square the side is s, the diagonal is $s\sqrt{2}$, and the area is s^2. The area of the triangle with altitude h is $\tfrac{1}{2}(s\sqrt{2})h = s^2$. $\therefore h = s\sqrt{2}$.

11. **(D)**

$$2^x = 8^{y+1} = 2^{3(y+1)}, \qquad \therefore x = 3y + 3;$$

$$3^{x-9} = 9^y = 3^{2y}, \qquad \therefore x - 9 = 2y.$$

The solution of the two linear equations is

$$x = 21, \quad y = 6; \qquad \therefore x + y = 27.$$

12. **(E)** To negate the statement "all members of a set have a given property" we write "some member of the set does not have the given property". Therefore, the negation is: For some x, $x^2 \not> 0$, that is, for some x, $x^2 \leq 0$.

13. **(A)** $(8 - r) + (13 - r) = 17$, $r = 2$, $s = 8 - r = 6$; $\therefore r:s = 1:3$.

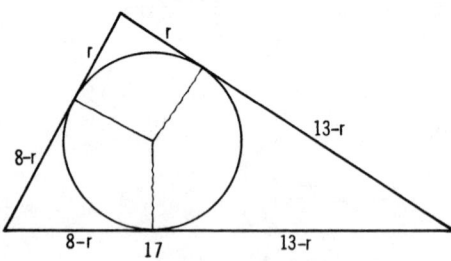

14. **(C)** Let P be the amount paid for the 749 sheep. Then, the selling price per sheep is $P/700$, and the complete profit is $49 \times P/700$. Therefore, the gain is

$$\frac{49 \times P/700}{P} \times 100 = 7 \text{ (per cent)}.$$

15. (B) Let the line cut the x-axis in $(-a, 0)$ and the y-axis in $(0, h)$. Then $\tfrac{1}{2}ah = T$ and $h = 2T/a$. The slope of the line is

$$\frac{h}{a} = \frac{2T}{a^2}$$

so its equation is

$$y = \frac{2T}{a^2}x + \frac{2T}{a} \quad \text{or} \quad 2Tx - a^2y + 2aT = 0.$$

16. (E) $x^2 + 3x + 2 = (x + 2)(x + 1)$. The integers s in S such that $s + 2$ is an integral multiple of 6 are $s = 4 + 6k$ ($k = 0, 1, 2, 3$); those such that $s + 1$ is an integral multiple of 6 are $s = 5 + 6k$ ($k = 0, 1, 2, 3$). Other possibilities are that $s + 2$ is a multiple of 2 and $s + 1$ a multiple of 3, or vice versa. In the first case $s = 2 + 6k$ ($k = 0, 1, 2, 3$), and in the second $s = 1 + 6k$ ($k = 0, 1, 2, 3, 4$). Thus the following 17 members of S lead to a zero remainder upon dividing $f(x)$ by 6: 4, 10, 16, 22; 5, 11, 17, 23; 2, 8, 14, 20; 1, 7, 13, 19, 25.

17. (D) x_1, y_1 cannot both be zero; for, if $x_1 = y_1 = 0$, points Q and R would not be distinct. Similarly, if $x_2 = y_2 = 0$, points P and R would not be distinct. If $y_1 = kx_1$, $y_2 = kx_2$ for some $k \neq 0$, the points P, Q, R lie on a straight line through O; otherwise an examination of the slopes of the segments shows that $OP \parallel QR$ and $OQ \parallel PR$, so the figure is a parallelogram.

OR

Either O, P, Q, R lie on a straight line, or the segments OR and PQ have distinct slopes and their midpoints are

$$\left(\frac{x_1 + x_2}{2}, \frac{y_1 + y_2}{2}\right), \quad \left(\frac{x_1 + x_2}{2}, \frac{y_1 + y_2}{2}\right)$$

respectively. But a quadrilateral whose diagonals bisect each other is a parallelogram.

18. (C) When two equations have the same graph, then every pair x, y satisfying one also satisfies the other. It follows that the coefficients of one are proportional to the corresponding coefficients of the other, so that $3k = c$, $kb = -2$, $kc = 12$. The solution of these three equations is $c = 6$, $b = -1$ or $c = -6$, $b = 1$.

19. (A)
$$2x - 3y = z, \quad x + 3y = 14z.$$
$$\therefore x = 5z, \quad y = 3z.$$
$$\therefore \frac{x^2 + 3xy}{y^2 + z^2} = \frac{x(x + 3y)}{y^2 + z^2} = \frac{5z \cdot 14z}{9z^2 + z^2} = \frac{70z^2}{10z^2} = 7.$$

20. (B) Since the expansion represents the binominal for all values of x and y, it represents the binomial for $x = 1$ and $y = 1$. For these values of x and y each term in the expansion has the same value as the numerical coefficient of the term. Therefore, the sum of the numerical coefficients of all the terms in the expansion equals $(1 - 2)^{18} = (-1)^{18} = 1$.

Part 2

21. (D) Let $\log_{b^2} x = m$ and let $\log_{x^2} b = n$. Then $x = b^{2m}$ and $b = x^{2n}$, so
$$(x^{2n})^{2m} = b^{2m}; \quad \therefore x^{4mn} = x,$$
$$\therefore 4mn = 1, \quad n = \frac{1}{4m}.$$
$$\therefore m + \frac{1}{4m} = 1 \text{ or } 4m^2 - 4m + 1 = 0. \quad \therefore m = \tfrac{1}{2} \text{ and } x = b.$$

22. (C) Since $DF = \tfrac{1}{3}DA$, area $(\triangle DFE) = \tfrac{1}{3}$ area $(\triangle DEA)$. Since E is the midpoint of DB,
$$\text{area }(\triangle DEA) = \tfrac{1}{2} \text{ area }(\triangle DBA).$$
Therefore, area $(\triangle DFE) = \tfrac{1}{3} \cdot \tfrac{1}{2}$ area $(\triangle DBA)$;
$$\therefore \text{area (quad. } ABEF) = \tfrac{5}{6} \text{ area }(\triangle DBA),$$
$$\therefore \text{area }(\triangle DFE) : \text{area (quad. } ABEF) = 1 : 5.$$

23. (D) Let the numbers be represented by x and y. Then
$$x + y = 7(x - y) \quad \text{and} \quad xy = 24(x - y).$$
Therefore,
$$8y = 6x \quad \text{and} \quad xy = 24x - 24y = 24x - 18x = 6x.$$
$\therefore x(y - 6) = 0$. The value $x = 0$ does not satisfy the conditions of the problem, so that $y = 6$. Since $6x = 8y$, $x = 8$. $\therefore xy = 48$.

24. (A) $y = (x-a)^2 + (x-b)^2 = 2[x^2 - (a+b)x] + a^2 + b^2$

$$= 2\left[x^2 - (a+b)x + \left(\frac{a+b}{2}\right)^2\right] + a^2 + b^2 - 2\left(\frac{a+b}{2}\right)^2$$

$$= 2\left[x - \frac{a+b}{2}\right]^2 + \frac{(a-b)^2}{2}.$$

For y to be a minimum, $x = (a+b)/2$.

OR

The graph of $y = 2x^2 - 2(a+b)x + a^2 + b^2$ is a parabola concave up with respect to the x-axis. It has a minimum point on the axis of symmetry, the equation of which is

$$x = -\frac{-2(a+b)}{4},$$

that is,

$$x = \frac{a+b}{2}.$$

25. (B) Let the linear factors be $x + ay + b$ and $x + cy + d$. When multiplied out and equated to the given expression, we obtain these equations between the coefficients: ① $a + c = 3$ ② $b + d = 1$ ③ $ad + bc = m$ ④ $bd = -m$ ⑤ $ac = 0$.
From equations ① and ⑤, either $c = 0$, $a = 3$ or $a = 0$, $c = 3$. With the first set of values equation ③ becomes $3d = m$. Substituting into equation ④, we have $3d = -bd$. If $d = 0$, $m = 0$. If $d \neq 0$, $b = -3$. ∴ $d = 4$ and $m = 12$. The second set of values, $a = 0$, $c = 3$, yields the same result because of the symmetry of the equations.

26. (C) From the given information the ratio of the distances covered by First, Second, and Third, in that order, is $10:8:6$. Since the speeds remain constant the ratio of the distances remains constant, so that, when Second completes an additional 2 miles, Third completes an additional x miles with

$$\frac{2}{x} = \frac{8}{6}, \quad x = 1\frac{1}{2}.$$

Therefore, when Second completes 10 miles, Third completes $7\frac{1}{2}$ miles, that is, Second beats Third by $2\frac{1}{2}$ miles.

27. (E) When $x \geq 4$, $|x-4|+|x-3| = x-4+x-3 \geq 1$.
When $x \leq 3$, $|x-4|+|x-3| = 4-x+3-x \geq 1$.
When $3 < x < 4$, $|x-4|+|x-3| = 4-x+x-3 = 1$.
$\therefore |x-4|+|x-3|$ is never less than 1. Since
$$a > |x-4|+|x-3|, \quad a > 1.$$

28. (D) Let a be the first term of the progression. Then
$$\frac{n}{2}[a + (a + 2(n-1))] = 153, \quad \text{or} \quad n^2 + n(a-1) - 153 = 0.$$

The product of the roots of this quadratic equation in n is -153. It suffices to list only integral factors of -153; for, if both roots are non-integral, neither yields a solution of our problem, and if one root is an integer, the other not, then their sum $a-1$ is not an integer, contrary to the assumption that a is an integer. Moreover, only positive integers greater than 1 are admissible for n. Therefore n may have any of the values 3, 9, 17, 51, 153.

29. (E) $\triangle RFD \sim \triangle SRF$ (an angle of one triangle equal to an angle of the other triangle and the including sides in proportion).

$$\therefore \frac{RS}{RF} = \frac{SF}{RD}, \quad \frac{RS}{5} = \frac{7\frac{1}{2}}{6}, \quad RS = 6\frac{1}{4}.$$

OR

By the law of cosines,
$$5^2 = 4^2 + 6^2 - 2\cdot 4\cdot 6 \cos \angle D, \quad \therefore \cos \angle D = \tfrac{27}{48} = \cos \angle RFS.$$
$$\therefore RS^2 = 5^2 + (7\tfrac{1}{2})^2 - 2(7\tfrac{1}{2})(5)(\tfrac{27}{48}), \quad \therefore RS = 6\tfrac{1}{4}.$$

30. (D) Let the roots be r and s with $r \geq s$.
$$\therefore r - s = \frac{[(2+\sqrt{3})^2 + 8(7+4\sqrt{3})]^{1/2}}{7+4\sqrt{3}}.$$

Since
$$7 + 4\sqrt{3} = (2+\sqrt{3})^2, \quad r - s = \frac{3}{2+\sqrt{3}} = 6 - 3\sqrt{3}.$$

OR

Let $r = 2 + \sqrt{3}$; since $7 + 4\sqrt{3} = (2+\sqrt{3})^2 = r^2$ we may write the equation as $r^2x^2 + rx - 2 = (rx - 1)(rx + 2) = 0$. Its roots are

$$\frac{1}{r} \quad \text{and} \quad -\frac{2}{r},$$

so that the larger root minus the smaller root is

$$\frac{1}{r} - \left(-\frac{2}{r}\right) = \frac{3}{r} = \frac{3}{2 + \sqrt{3}} = 6 - 3\sqrt{3}.$$

Part 3

31. (B) $f(n+1) - f(n-1)$

$$= \frac{5 + 3(5)^{1/2}}{2}\left(\frac{1 + (5)^{1/2}}{2}\right)^{n+1} + \frac{5 - 3(5)^{1/2}}{2}\left(\frac{1 - (5)^{1/2}}{2}\right)^{n+1}$$

$$- \frac{5 + 3(5)^{1/2}}{2}\left(\frac{1 + (5)^{1/2}}{2}\right)^{n-1} - \frac{5 - 3(5)^{1/2}}{2}\left(\frac{1 - (5)^{1/2}}{2}\right)^{n-1}$$

$$= \frac{5 + 3(5)^{1/2}}{2}\left(\frac{1 + (5)^{1/2}}{2}\right)^{n}\left[\frac{1 + (5)^{1/2}}{2} - \frac{2}{1 + (5)^{1/2}}\right]$$

$$+ \frac{5 - 3(5)^{1/2}}{2}\left(\frac{1 - (5)^{1/2}}{2}\right)^{n}\left[\frac{1 - (5)^{1/2}}{2} - \frac{2}{1 - (5)^{1/2}}\right]$$

$$= \frac{5 + 3(5)^{1/2}}{2}\left(\frac{1 + (5)^{1/2}}{2}\right)^{n}(1)$$

$$+ \frac{5 - 3(5)^{1/2}}{2}\left(\frac{1 - (5)^{1/2}}{2}\right)^{n}(1) = f(n).$$

32. (C) Since

$$\frac{a+b}{b+c} = \frac{c+d}{d+a}, \quad \text{then} \quad \frac{a+b}{c+d} = \frac{b+c}{d+a}$$

and

$$\frac{a+b}{c+d} + 1 = \frac{b+c}{d+a} + 1.$$

$$\therefore \frac{a+b+c+d}{c+d} = \frac{a+b+c+d}{a+d}.$$

If $a+b+c+d \neq 0$, then $a = c$. If $a+b+c+d = 0$, then a may or may not equal c, so that the correct choice is (C).

OR

If
$$\frac{a+b}{b+c} = \frac{c+d}{d+a},$$
then $(a+b)(a+d) = (c+d)(c+b)$, that is
$$a^2 + (b+d)a = c^2 + (b+d)c.$$
Hence
$$a^2 - c^2 + (b+d)(a-c) = 0, \quad \therefore (a-c)(a+c+b+d) = 0.$$

33. (B) Draw a line through $P \parallel$ to one of the sides of the rectangle; see figure.
$$16 - c^2 = 9 - d^2$$
$$25 - c^2 = x^2 - d^2 \quad \therefore x^2 - 9 = 9 \quad \therefore x = 3\sqrt{2}.$$

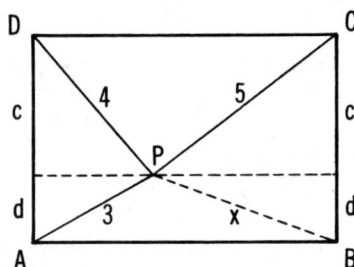

34. (C)
$$s = 1 + 2i + 3i^2 + \cdots + (n+1)i^n$$
$$is = i + 2i^2 + \cdots + ni^n + (n+1)i^{n+1}$$
$$s(1-i) = 1 + i + i^2 + \cdots + i^n - (n+1)i^{n+1}$$
$$s(1-i) = \frac{1 - i^{n+1}}{1-i} - (n+1)i^{n+1} = 1 - (n+1)i$$

since $i^n = 1$ if n is a multiple of 4.
$$\therefore s = \frac{1-(n+1)i}{1-i} \cdot \frac{1+i}{1+i} = \frac{1}{2}(n+2-ni).$$

SOLUTIONS: 1964 EXAMINATION   93

OR

$$s = 1 + 3i^2 + 5i^4 + 7i^6 + \cdots + (n+1)i^n$$
$$\qquad\qquad\qquad + 2i + 4i^3 + \cdots + ni^{n-1}$$
$$s = (1-3) + (5-7) + \cdots + ((n-3)-(n-1))$$
$$\qquad + n + 1 + i[(2-4) + (6-8) + \cdots + (n-2-n)]$$
$$s = \frac{n}{4}(-2) + n + 1 + \frac{n}{4}(-2)i = \tfrac{1}{2}(n+2-ni).$$

35. (B) $14^2 - q^2 = 13^2 - p^2$, $27 = q^2 - p^2$, $15 = q + p$ ∴ $q = \frac{42}{5}$, $p = \frac{33}{5}$.
$13^2 - r^2 = 15^2 - s^2$, $56 = s^2 - r^2$, $14 = s + r$ ∴ $s = 9$, $r = 5$.
$AD^2 = 13^2 - 5^2$, $AD = 12$, $BE^2 = 13^2 - (\frac{33}{5})^2$, $BE = \frac{56}{5}$.
△HDB ~ △HEA;

$$\therefore \frac{t}{u} = \frac{r}{p} = \frac{\frac{56}{5} - u}{12 - t} = \frac{HB}{HA}.$$

Since $r = 5$, $p = \frac{33}{5}$, $u = \frac{33}{25}t$,

$$12 - t = \frac{33 \cdot 56}{125} - \frac{33 \cdot 33}{25 \cdot 25}t.$$

∴ $t = \frac{435}{116}$, $12 - t = \frac{957}{116}$. ∴ $HD:HA = 435:957 = 5:11$.

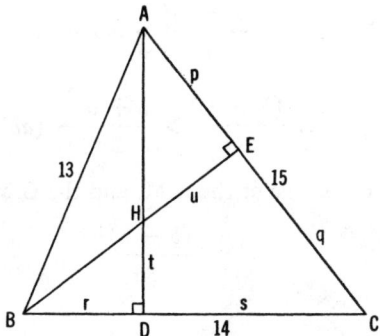

36. (E) Diagram I on p. 94 shows the circle with its center at the vertex C of the equilateral triangle. In this position, the arc in question is 60° since it is subtended by a central angle of 60°. The second diagram shows the circle in another permissible position. Its center lies on a line parallel to the base of the triangle. Extend line NC to meet the circle again at D. The diameter through C bisects

the vertex angle of isosceles triangle MCD. Therefore, the peripheral angle MDN is $30°$ and the subtended arc MN is $60°$.

I

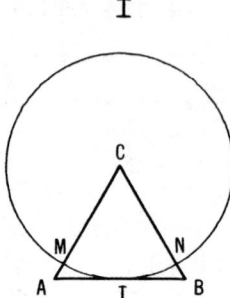

II

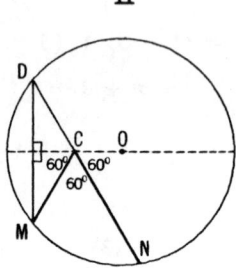

37. (D) We first prove that the A.M., $\frac{1}{2}(b + a)$, is greater than the G.M., $(ab)^{1/2}$, for $b \neq a$.

$$(b - a)^2 = b^2 - 2ab + a^2 > 0, \qquad b^2 + 2ab + a^2 > 4ab,$$

$$b + a > 2(ab)^{1/2}, \qquad \frac{b + a}{2} > (ab)^{1/2}.$$

$$\therefore \frac{b + a}{2} - a > (ab)^{1/2} - a, \qquad \frac{b - a}{2} > a^{1/2}(b^{1/2} - a^{1/2}).$$

Therefore, if

$$b > a, \qquad \frac{(b - a)^2}{4} > a[b + a - 2(ab)^{1/2}],$$

$$\therefore \frac{(b - a)^2}{8a} > \frac{b + a}{2} - (ab)^{1/2},$$

that is, the difference of the A.M. and the G.M. is less than

$$\frac{(b - a)^2}{8a}.$$

38. (D) Let y denote half the length of QR, and let x be the distance from M to the foot of the altitude from P. Then the square of the altitude may be written

$$16 - (y - x)^2 = (\tfrac{7}{2})^2 - x^2,$$

$$\therefore (y - x)^2 - x^2 = \tfrac{15}{4} = y^2 - 2xy.$$

$$16 - (y - x)^2 = 7^2 - (y + x)^2,$$

SOLUTIONS: 1964 EXAMINATION   95

$$\therefore (y+x)^2 - (y-x)^2 = 33 = 4xy,$$

and $2xy = \frac{33}{2}$ $\therefore y^2 - \frac{33}{2} = \frac{15}{4}$, $\therefore y^2 = \frac{81}{4}$, $y = \frac{9}{2}$. $\therefore QR = 9$.

OR

By the law of cosines, $(\frac{7}{2})^2 + y^2 - 2 \cdot \frac{7}{2} \cdot y \cos \angle PMQ = 16$ and
$(\frac{7}{2})^2 + y^2 - 2 \cdot \frac{7}{2} \cdot y \cos (180° - \angle PMQ)$
$$= (\frac{7}{2})^2 + y^2 + 7y \cos \angle PMQ = 49.$$
$\therefore \frac{49}{2} + 2y^2 = 65$, $y = \frac{9}{2}$, $2y = 9$.

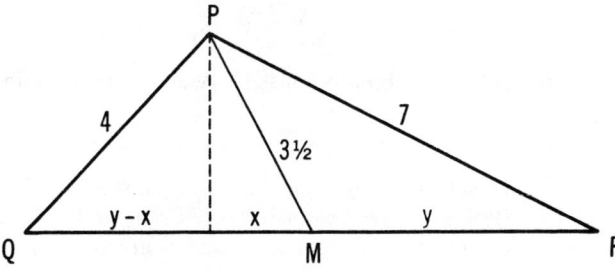

39. (A) The length of each segment through P is less than the length of the larger of the two sides of the triangle from the same vertex as the segment under consideration. This fact is a consequence of the theorem: in a triangle, the larger of two sides lies opposite the larger of two angles. Thus $AA' + BB' + CC' < b + a + a$, and so $s < 2a + b$. To see that each of the other alternatives may be violated, consider a point P very close to one of the end-points of the longest side.

40. (A) First we note that the number of minutes in a day is $24 \times 60 = 1440$, while the number of minutes as recorded by the watch in a day is $2\frac{1}{2}$ less or $1437\frac{1}{2}$, so that the correction factor for this watch is $1440/1437\frac{1}{2}$ or, more simply, $576/575$.
Designate a minute recorded according to the watch as a "watch-minute" and a day recorded by the watch as a "watch-day". Since the time interval given is $5\frac{5}{6}$ watch-days, or $5\frac{5}{6} \times 24 \times 60$ watch-minutes, we have $5\frac{5}{6} \times 24 \times 60 + n = 5\frac{5}{6} \times 24 \times 60 \times \frac{576}{575}$.
$\therefore n = 140 \cdot 60 (\frac{576}{575} - 1) = 14\frac{14}{23}$.

1965 Solutions

Part 1

1. (C) $2^{2x^2-7x+5} = 1 = 2^0$, $\therefore 2x^2 - 7x + 5 = 0$. Since $7^2 - 4\cdot 2\cdot 5 = 9 > 0$, this equation has two real solutions.

2. (D) Let r be the magnitude of the radius of the circle; then the length of a side of the hexagon is r and the length of the shorter arc intercepted by the side is $2\pi r/6$. The required ratio is $r:(2\pi r/6)$ or $3:\pi$.

3. (B) $$81^{-(2^{-2})} = 81^{-1/4} = \frac{1}{81^{1/4}} = \frac{1}{3}.$$

   Query: Is there another permissible value not listed in the five choices?

4. (C) Let l_4 be the set of points such that each point is equidistant from l_1 and l_3. Designate the distance by d. Let l_5 and l_6 be the set of points at the distance d from l_2 (l_5 and l_6 are parallel to l_2). The intersection set of l_4, l_5, and l_6, containing two points, is the required set of points.

5. (A) $$0.363636\cdots = \frac{36}{10^2} + \frac{36}{10^4} + \cdots = \frac{36/10^2}{1 - 1/10^2} = \frac{36}{99} = \frac{4}{11},$$

   so that the correct answer is (A).

   OR

   Let
   $$F = 0.363636\cdots, \qquad 100F = 36.363636\cdots,$$
   $$99F = 36, \qquad F = \frac{36}{99} = \frac{4}{11}.$$

6. (B) Since $10^{\log_{10} 9} = 9$, $8x + 5 = 9$, $\therefore x = 1/2$.

7. (E) Let the roots be r and s; then $r + s = -b/a$ and $rs = c/a$.
   $$\therefore \frac{1}{r} + \frac{1}{s} = \frac{r+s}{rs} = \frac{-b/a}{c/a} = -b/c, \qquad c \neq 0, \qquad a \neq 0.$$

8. (A) Let s be the length of the required segment; then $s^2/18^2 = 2/3$ so that $s = 6(6)^{1/2}$ (the areas of two similar triangles are to each other as the squares of corresponding sides).

9. (E) $y = x^2 - 8x + c = x^2 - 8x + 16 + c - 16 = (x - 4)^2 + c - 16$ (a parabola). For the vertex to be on the x-axis, its coordinates must be $(4, 0)$ so that c must have the value 16.

OR

Since the x-coordinate of the vertex of a parabola $y = ax^2 + bx + c$ is $-b/2a$, we have, in this instance,

$$x = -\frac{-8}{2} = 4.$$

The y-coordinate is 0 if $0 = 4^2 - 8 \cdot 4 + c$, i.e., if $c = 16$.

10. (A) $\qquad x^2 - x - 6 < 0, \qquad x^2 - x < 6,$

$\qquad x^2 - x + \frac{1}{4} < 6 + \frac{1}{4}, \qquad (x - \frac{1}{2})^2 < (\frac{5}{2})^2.$

$\therefore |x - \frac{1}{2}| < \frac{5}{2}$ or $-\frac{5}{2} < x - \frac{1}{2} < \frac{5}{2}$, that is, $-2 < x < 3$.

OR

$x^2 - x - 6 < 0$, $(x - 3)(x + 2) < 0$. This inequality is satisfied if $x - 3 < 0$ and $x + 2 > 0$ or if $x - 3 > 0$ and $x + 2 < 0$. The first set of inequalities implies $-2 < x < 3$; the second set is impossible to satisfy.

11. (B) $\qquad (-4)^{1/2} = 2(-1)^{1/2}, \qquad (-16)^{1/2} = 4(-1)^{1/2}.$

$\therefore (-4)^{1/2}(-16)^{1/2} = 8(-1) = -8$

but

$[(-4)(-16)]^{1/2} = (64)^{1/2} = 8.$

$\therefore$ Statement I is incorrect.

12. (D) Since $\triangle BDE \sim \triangle BAC$, $s/6 = (12 - s)/12$, $s = 4$.

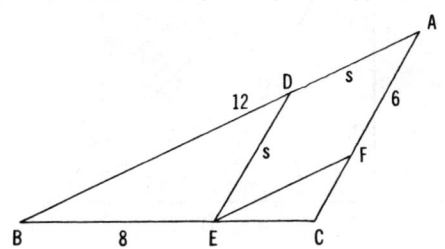

13. **(E)** We must find out if the line intersects the circle $x^2 + y^2 = 16$ in two points R and S, or if it is tangent to the circle, or if it has no points in common with the circle. It is easy to see that the line $5y - 3x = 15$ intersects $x^2 + y^2 = 16$ in two points because, for example, the point $(0, 3)$ on the line is in the interior of the circle (since $0^2 + 3^2 < 16$), so a whole segment RS of the line is inside the circle. This segment has infinitely many points x, y and the coordinates of each satisfy $5y - 3x = 15$ and $x^2 + y^2 \leq 16$.

14. **(A)** $(x^2 - 2xy + y^2)^7 = ((x-y)^2)^7 = (x-y)^{14}$. By setting $x = y = 1$ in the expansion of this binomial we obtain the sum s of the integer coefficients. Consequently $s = (1-1)^{14} = 0$.

15. **(B)** $\qquad 52_b = 5b + 2, \qquad 25_b = 2b + 5.$

    Since $\qquad 52_b = 2(25_b), \qquad 5b + 2 = 2(2b + 5); \quad \therefore b = 8.$

16. **(C)** Draw AE and the altitude FG to the base DE of triangle DEF. Since F is the intersection point of the medians of triangle ACE, $FD = \frac{1}{3}AD$. $\therefore FG = \frac{1}{3}AC = \frac{1}{3} \cdot 30 = 10$.

    $\therefore$ area $(\triangle DEF) = \frac{1}{2} \cdot 15 \cdot 10 = 75$.

    OR

    The three medians of a triangle subdivide it into six triangles of equal area. Therefore

    Area $(\triangle FDE) = \frac{1}{6} \cdot \frac{1}{2} \cdot 30 \cdot 30 = \frac{450}{6} = 75$.

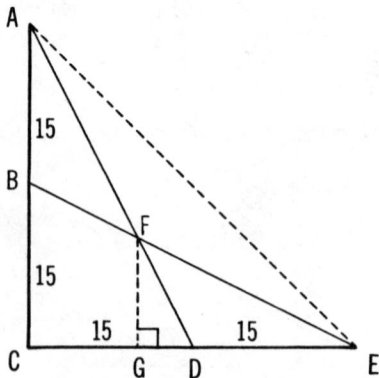

SOLUTIONS: 1965 EXAMINATION   99

17. (E) The given statement may be rephrased as "If the picnic on Sunday is not held, then the weather is not fair." Its contrapositive is "If the weather is fair, then the picnic is held on Sunday." This is statement (E).

18. (B) The error in using the approximation $1 - y$ is

$$\frac{1}{1+y} - (1-y) = \frac{y^2}{1+y}.$$

The ratio of the error to the correct value is, therefore,

$$\left(\frac{y^2}{1+y}\right) \Big/ \left(\frac{1}{1+y}\right) = y^2.$$

19. (C) Let

$$x^4 + 4x^3 + 6px^2 + 4qx + r = (x+a)(x^3 + 3x^2 + 9x + 3)$$
$$= x^4 + (a+3)x^3 + (3a+9)x^2 + (9a+3)x + 3a.$$

Equating the coefficients of like powers of x, we have $a = 1$, $p = 2$, $q = 3$, $r = 3$, so that $(p+q)r = (2+3)(3) = 15$.

OR

Divide $x^4 + 4x^3 + 6px^2 + 4qx + r$ by $x^3 + 3x^2 + 9x + 3$. The quotient is $x + 1$ and the remainder is the second-degree polynomial $(6p - 12)x^2 + (4q - 12)x + r - 3$. This remainder equals zero since the division is exact. Therefore $p = 2$, $q = 3$, $r = 3$ and $(p+q)r = 15$.

20. (C) The rth term equals $S_r - S_{r-1}$, and $S_r = 2r + 3r^2$, and

$$S_{r-1} = 2(r-1) + 3(r-1)^2 = 3r^2 - 4r + 1.$$

∴ the rth term equals $6r - 1$.

OR

Let the rth term be u_r and let the first term be a; then

$$S_r = (r/2)(a + u_r) = 2r + 3r^2, \quad \therefore a + u_r = 4 + 6r.$$

But

$$a = S_1 = 2 \cdot 1 + 3 \cdot 1^2 = 5. \quad \therefore u_r = 4 + 6r - 5 = 6r - 1.$$

OR

$a = S_1 = 5$, $S_2 = 16$. ∴ $u_2 = S_2 - S_1 = 11$, but $u_2 = a + d$;
∴ $d = 6$. ∴ $u_r = a + (r-1)d = 5 + (r-1)(6) = 6r - 1$.

Part 2

21. (D)

$$\log_{10}(x^2 + 3) - 2\log_{10} x = \log_{10}\frac{x^2+3}{x^2} = \log_{10}\left(1 + \frac{3}{x^2}\right).$$

For a sufficiently large value of x, $3/x^2$ may be made less than a specified positive number N and so $\log_{10}(1 + 3/x^2)$ may be made less than the specified positive number $\log_{10}(1 + N)$. Challenge: If, in the problem, the condition $x > \frac{1}{2}$ were given instead of $x > \frac{2}{3}$, show that then (C) would also be an acceptable answer.

22. (A)

$$a_2 x^2 + a_1 x + a_0 = a_2\left(x^2 + \frac{a_1}{a_2}x + \frac{a_0}{a_2}\right)$$
$$= a_2(x^2 - (r+s)x + rs)$$
$$= a_2(r-x)(s-x)$$

If $rs = a_0/a_2$ is not zero, then $a_0 \neq 0$ and the last expression may be written in the form

$$a_2 rs\left(1 - \frac{x}{r}\right)\left(1 - \frac{x}{s}\right).$$

$$\therefore a_2 x^2 + a_1 x + a_0 = a_2\left(\frac{a_0}{a_2}\right)\left(1 - \frac{x}{r}\right)\left(1 - \frac{x}{s}\right) = a_0\left(1 - \frac{x}{r}\right)\left(1 - \frac{x}{s}\right)$$

for all values of x provided $a_0 \neq 0$.

23. (D) Since $|x - 2| < 0.01$, x is positive and $x < 2.01$. Hence
$$|x^2 - 4| = |x - 2||x + 2|$$
$$= |x - 2|(x + 2) < (.01)(4.01) = .0401.$$

OR

$|x - 2| < 0.01$ implies $1.99 < x < 2.01$.

$$\therefore 3.9599 < 3.9601 < x^2 < 4.0401.$$

$\therefore -.0401 < x^2 - 4 < .0401$, that is, $|x^2 - 4| < .0401$.

24. (E) Let $P = 10^{1/11} \cdot 10^{2/11} \cdots 10^{n/11} = 10^s$ where

$$s = \frac{1 + 2 + \cdots + n}{11} = \frac{1}{11} \cdot \frac{1}{2} n(n+1).$$

SOLUTIONS: 1965 EXAMINATION

For $P > 100000$, $10^s > 10^5$, that is, $s > 5$.

$$\therefore \frac{n^2 + n}{22} > 5, \quad n^2 + n - 110 > 0,$$

$$(n + 11)(n - 10) > 0, \quad n > 10.$$

25. (D) Since CB is the median to the hypotenuse AE of right triangle AEC, $CB = \frac{1}{2}AE = AB$.

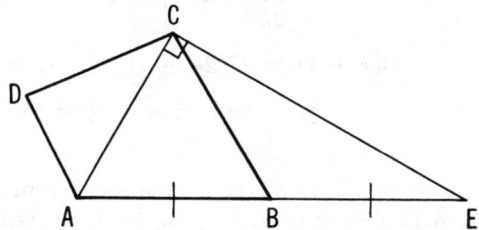

26. (E) From the given information, $2k = a + b$, $3l = c + d + e$, so

$$m = \frac{a + b + c + d + e}{5} = \frac{2k + 3l}{5} = \frac{4k + 6l}{10},$$

while

$$p = \frac{k + l}{2} = \frac{5k + 5l}{10}.$$

Therefore $m >, =, < p$ according as $4k + 6l >, =, < 5k + 5l$, that is, according as $l > k$, $l = k$ or $l < k$.

Since k depends only on the first two of the numbers a, b, c, d, e, while l depends only on the last three, all three possibilities can be realized by proper choice of the five numbers.

27. (A) Since $y^2 + my + 2 = (y - 1)f(y) + R_1$ for all values of y, we have, letting $y = 1$, $3 + m = R_1$. Similarly, since

$$y^2 + my + 2 = (y + 1)g(y) + R_2$$

for all values of y, we have, letting $y = -1$, $3 - m = R_2$. Since $R_1 = R_2$, $3 + m = 3 - m$. $\therefore m = 0$.

28. (B) Let the time needed for a full escalator step to reach the next level be designated as unit time, and take the visible length of the escalator to be n units of distance. Suppose Z takes k steps per unit time; then one of Z's steps takes $1/k$ units of time, and 18

steps take $18/k$ units of time. Z's rate of descent is $k + 1$ units of distance per unit of time. Therefore,

$$\frac{18}{k}(k + 1) = n = \frac{36}{2k}(k + 1).$$

A takes $2k$ steps per unit of time, so his 27 steps take $27/2k$ units of time, and his rate of descent is $2k + 1$ units of distance per unit of time. Hence

$$\frac{27}{2k}(2k + 1) = n.$$

Thus $36(k + 1) = 27(2k + 1), \quad 18k = 9,$

$k = \frac{1}{2}, \quad$ and $\quad n = 36 \cdot \frac{3}{2} = 54.$

29. (A) Let n_1 ($n_1 \geq 0$) represent the number of students taking Mathematics and English only. Let n_2 ($n_2 > 0$, n_2 even) represent the number of students taking all three subjects.
From the given information we have, then,

$$n_1 + n_1 + 5n_2 + n_2 + 6 = 28, \quad \text{that is,} \quad n_1 + 3n_2 = 11.$$

Under the restrictions given in the problem, this equation implies $n_2 = 2$ and $n_1 = 5$.

30. (B) $\angle A = \angle BCD$ (Why?), $DF = CF$ (Why?). $\therefore \angle DCF = \angle CDF.$

$\angle BCD = 90° - \angle DCF, \quad \angle FDA = 90° - \angle CDF,$

$\therefore \angle FDA = \angle BCD = \angle A.$ $\therefore DF = FA = CF$, that is, DF bisects CA. Also $\angle CFD = \angle FDA + \angle A = 2 \angle A.$

The given information is, therefore, sufficient to prove choices (A), (C), (D), and (E). For choice (B) to be true, segments CD and AD would have to be equal, but such equality can not hold for all possible positions of point D.

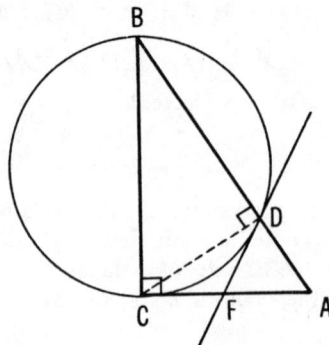

SOLUTIONS: 1965 EXAMINATION   103

OR

Draw the radius OD. If right $\triangle BDC$ is turned counterclockwise by 90° about the point D, corresponding sides DB and DC, and corresponding sides DC and DA of similar triangles BDC and CDA will coincide. Since $OD \perp DF$, these lines will also coincide after the rotation Therefore DF is a median to the hypotenuse of $\triangle CDA$, so $CF = FA = DF$. $\angle CFD$ is an exterior angle of isosceles $\triangle FDA$, hence equal to $2 \angle A$. However, the median to the hypotenuse of a right triangle bisects the right angle only in the special case that the legs are equal.

Part 3

31. (C) Let $\log_a x = y$; $\therefore x = a^y$, $\therefore \log_b x = \log_b a^y = y \log_b a$.

$\therefore (\log_a x)(\log_b x) = \log_a b$ implies $y \cdot y \log_b a = \log_a b$.

Since $\log_b a = 1/\log_a b$, $y^2 = (\log_a b)^2$, $y = \log_a b$ or $y = -\log_a b$;
$\therefore \log_a x = \log_a b$ or $\log_a x = \log_a b^{-1}$. $\therefore x = b$ or $x = b^{-1} = 1/b$.

OR

Let

$\log_b x = y$; $\therefore x = b^y$. $\therefore (\log_a b^y)(\log_b b^y) = \log_a b$,

$(y \log_a b)(y \log_b b) = \log_a b$, $y^2 \log_a b = \log_a b$;

$\therefore y^2 = 1$, $\therefore y = 1$ or $y = -1$, $\therefore x = b$ or $x = b^{-1}$.

OR

$(\log_b x)(\log_a x) = \log_a x^{(\log_b x)} = \log_a b$. $\therefore x^{\log_b x} = b$.

$\therefore (\log_b x)(\log_b x) = \log_b b = 1$. $\therefore \log_b x = 1$ or $\log_b x = -1$.
$\therefore x = b$ or $x = b^{-1}$.

32. (C) $100 = C - x$, $C = 100 + x$.

$S' = C + 1\frac{1}{9} = 100 + x + \frac{10}{9} = 100 + \frac{x}{100}\left(100 + x + \frac{10}{9}\right)$,

$\therefore \frac{x^2}{100} + \frac{x}{90} - \frac{10}{9} = 0$, $x = 10$.

33. (D) $15! = 10^3(3 \cdot 14 \cdot 13 \cdot 12 \cdot 11 \cdot 9 \cdot 8 \cdot 7 \cdot 6 \cdot 3 \cdot 2 \cdot 1)$ and

$15! = 12^5(15 \cdot 14 \cdot 13 \cdot 11 \cdot 5 \cdot 7 \cdot 5)$.

$\therefore k = 5$ and $h = 3$. $\therefore k + h = 8$.

104   THE MAA PROBLEM BOOK II

34. (B) We seek the smallest value of

$$y = \frac{4x^2 + 8x + 13}{6(x+1)} = \frac{4(x^2 + 2x + 1) + 9}{6(x+1)} = \frac{4(x+1)^2 + 9}{6(x+1)}$$

under the condition $x \geq 0$. Let $x + 1 = z$; then the expression whose minimum we must determine may be written

$$y = \frac{4z^2 + 9}{6z}, \quad \text{and since} \quad x \geq 0, \quad z \geq 1.$$

We multiply both sides by $6z$ and obtain

$$4z^2 + 9 = 6zy \quad \text{or} \quad 4z^2 - 6zy + 9 = 0.$$

We write z in terms of y:

$$z = \frac{6y}{8} \pm \frac{1}{8}(36y^2 - 144)^{1/2} = \frac{3y}{4} \pm \frac{3}{4}(y^2 - 4)^{1/2}.$$

Since z is real, $y^2 \geq 4$, so $y \geq 2$ or $y \leq -2$. However, the condition $x \geq 0$ implies that the expression y is never negative. Hence, the smallest value y may have is 2. When

$$y = 2, \quad z = \frac{6 \cdot 2}{8} = \frac{3}{2}, \quad \text{so} \quad x = \frac{1}{2}.$$

35. (D) Suppose the rectangle has been folded so that the crease is along EF and AE lies along EC; see figure. Let $EB = x$; then

$$AE = EC = 5 - x \quad \text{and} \quad DF = x,$$

so that $AG = x$ and $GE = 5 - 2x$. $\therefore w^2 = (6^{1/2})^2 - (5 - 2x)^2$ and, also, $w^2 = (5 - x)^2 - x^2$. $\therefore 6 - 25 + 20x - 4x^2 = 25 - 10x$. $\therefore x = 2$ and $w = 5^{1/2}$. The value $x = 11/2$ is rejected.

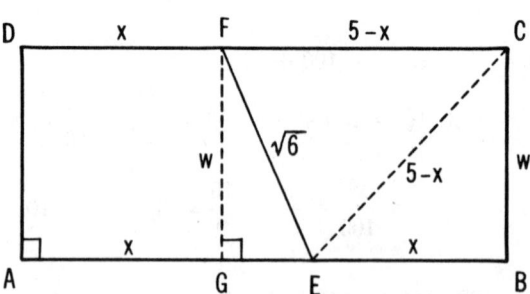

36. (E) The right triangles PP_1P_2, $P_1P_2P_3$, $P_2P_3P_4$, $\cdots$ are all similar;

SOLUTIONS: 1965 EXAMINATION   105

hence
$$\frac{a}{b} = \frac{b}{P_2 P_3} = \frac{P_2 P_3}{P_3 P_4} = \cdots,$$

and $\quad P_2 P_3 = \dfrac{b^2}{a}, \quad P_3 P_4 = \dfrac{b^3}{a^2}, \quad \cdots, \quad P_n P_{n+1} = \dfrac{b^n}{a^{n-1}}.$

The sum of the lengths of these segments is

$$a + b + \frac{b^2}{a} + \frac{b^3}{a^2} + \cdots = a\left[1 + \frac{b}{a} + \left(\frac{b}{a}\right)^2 + \cdots\right]$$

$$= a \cdot \frac{1}{1 - (b/a)} = \frac{a^2}{a - b}$$

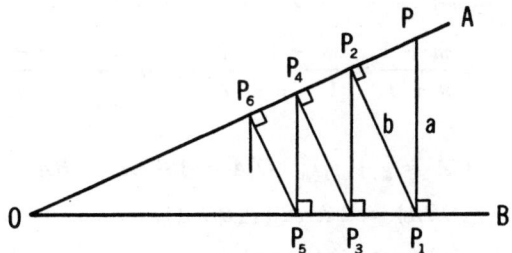

37. (C) Draw $DGH \parallel AB$. $\therefore DG:3a = b:3b$; $DG = a = EA$. $\therefore EF = FG$ and $AF = FD$, so that $AF/FD = 1$. Also $DH:4a = b:3b$, $DH = 4a/3$ and $GH = DH - DG = a/3$; $\therefore GC = \tfrac{1}{3}EC$ and $EG = \tfrac{2}{3}EC$, and, since $EF = FG$, $FC = \tfrac{2}{3}EC$. $\therefore EF/FC = \tfrac{1}{2}$. $\therefore (EF/FC) + (AF/FD) = \tfrac{1}{2} + 1 = \tfrac{3}{2}$.

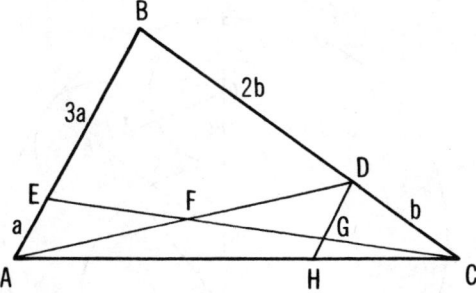

38. (E) Denote by a, b, and c the respective times it takes A, B, and C to do the work alone. Then

$$m \cdot \frac{1}{a} = \frac{1}{b} + \frac{1}{c}, \quad n \cdot \frac{1}{b} = \frac{1}{a} + \frac{1}{c}, \quad x \cdot \frac{1}{c} = \frac{1}{a} + \frac{1}{b}.$$

Hence
$$\frac{m}{a} - \frac{n}{b} = \frac{1}{b} - \frac{1}{a}, \quad \frac{1}{a}(m+1) = \frac{1}{b}(n+1), \quad \text{so} \quad \frac{a}{b} = \frac{m+1}{n+1}.$$

Also,
$$\frac{m}{a} + \frac{n}{b} = \frac{1}{b} + \frac{1}{a} + \frac{2}{c} = \frac{1}{b} + \frac{1}{a} + \frac{2}{x}\left(\frac{1}{a} + \frac{1}{b}\right),$$

$$\frac{1}{a}\left(m - 1 - \frac{2}{x}\right) = \frac{1}{b}\left(1 - n + \frac{2}{x}\right),$$

so
$$\frac{a}{b} = \frac{m - 1 - 2/x}{1 - n + 2/x}.$$

$$\therefore \frac{m+1}{n+1} = \frac{m - 1 - 2/x}{1 - n + 2/x}, \quad \text{so} \quad x = \frac{m+n+2}{mn-1}.$$

39. (B) $OA = \frac{1}{2} + r, \quad O'A = 1 + r, \quad BA = \frac{3}{2} - r.$

Area $(\triangle OBA) = 2$ Area $(\triangle BO'A)$.

Semiperimeter $(\triangle OBA) = \frac{1}{2}(\frac{1}{2} + r + 1 + \frac{3}{2} - r) = \frac{3}{2}$.

Semiperimeter $(\triangle BO'A) = \frac{1}{2}(\frac{3}{2} - r + \frac{1}{2} + 1 + r) = \frac{3}{2}$.

$\therefore [\frac{3}{2}(1-r)(\frac{1}{2})r]^{1/2} = 2[\frac{3}{2}(r)(1)(\frac{1}{2}-r)]^{1/2}.$

$\therefore 7r = 3, \quad r = \frac{3}{7}, \quad d = \frac{6}{7} \approx .86.$

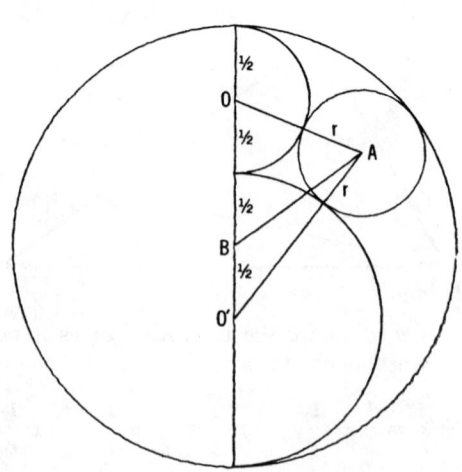

SOLUTIONS: 1965 EXAMINATION

OR

Draw the altitude h from A to OB. We have
$$(\tfrac{1}{2}+r)^2 - t^2 = (\tfrac{3}{2}-r)^2 - (1-t)^2,$$
$$t = 2r - \tfrac{1}{2}, \qquad h^2 = 3r - 3r^2.$$

Since
$$\text{Area }(\triangle OAO') = \tfrac{1}{2}h \cdot \tfrac{3}{2} = [(\tfrac{3}{2}+r)(r)(1)(\tfrac{1}{2})]^{1/2},$$

we have
$$\tfrac{9}{16}h^2 = \tfrac{9}{16}(3r - 3r^2) = \tfrac{3}{4}r + \tfrac{1}{2}r^2 \quad \therefore r = \tfrac{3}{7} \text{ and } d = \tfrac{6}{7}.$$

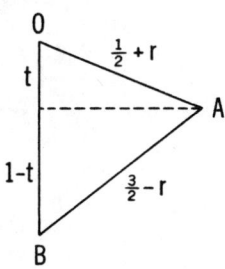

40. **(D)** Let
$$P = x^4 + 6x^3 + 11x^2 + 3x + 31$$
$$= (x^2 + 3x + 1)^2 - 3(x - 10) = y^2.$$

When $x = 10$, $P = (x^2 + 3x + 1)^2 = 131^2 = y^2$. To prove that 10 is the only possible value we use the following lemma: If
$$|N| > |M|,$$
N, M integers, then
$$N^2 - M^2 \geq 2|N| - 1.$$

(This lemma is easy to prove; try it.)

Case I If $x > 10$, then
$$3(x - 10) = (x^2 + 3x + 1)^2 - y^2 \geq 2|x^2 + 3x + 1| - 1,$$
an impossibility.

Case II If $x < 10$, then
$$3(10 - x) = y^2 - (x^2 + 3x + 1)^2$$
$$\geq 2|y| - 1 > 2|x^2 + 3x + 1| - 1.$$

This inequality holds for the integers
$$x = 2, 1, 0, -1, -2, -3, -4, -5, -6,$$
but none of these values makes P the square of an integer.

IV

Classification of Problems

To classify these problems is not a simple task; their content is so varied and their solution-possibilities so diverse that it is difficult to pigeonhole them into a few categories. Moreover, no matter which headings are selected, there are borderline cases that need cross-indexing.

Arithmetic, Algebra, Geometry, and *Miscellaneous Topics* constitute the primary groupings. A more refined classification is provided by the subheadings. Subsumed under *Arithmetic* are approximate-number and number-theoretic problems, as well as problems on means, percentage, and profit-and-loss; these are also cross-indexed for the appropriate subheadings under *Algebra.* Under *Geometry* are listed the problems on Euclidean geometry and simple coordinate geometry.

The number preceding the semicolon refers to the last two digits of the examination year, and the numbers following the semicolon refer to the problems in that examination. For example, 61; 13 means Problem 13 in the 1961 examination.

Arithmetic

Approximation	65; 18
Mean	62; 8, 10 63; 11 64; 37 65; 26
Number theory	61; 17, 28, 35 62; 22 63; 8, 13 64; 3 65; 15, 33, 40
Percent	61; 18 64; 14 65; 32
Profit and loss	64; 9, 14 65; 32
Proportion	see under Algebra

Algebra

Binomial expansion	see Polynomial expansion and Progression
Discriminants	see Roots
Equations	
Cubic (and higher)	62; 39
Indeterminate	61; 33 62; 36 63; 31 64; 11, 29, 40
Linear	62; 10, 24, 35 63; 36 64; 26, 40 65; 28, 29, 37, 38
Quadratic	61; 29 62; 34, 38 63; 3, 14, 20, 24, 28, 29 64; 7, 8, 30 65; 1, 22, 24, 27, 34, 35, 39
Radical	63; 40 65; 39
Systems	63; 4, 23 64; 18, 23, 25 65; 29, 38
Exponents	61; 1, 9, 28 62; 1, 4 63; 2 65; 1, 3, 33
Extreme value	61; 34, 40 62; 26, 37 63; 19, 28, 29, 37 64; 24 65; 21, 34
Factors	61; 22 62; 9 63; 21 64; 25 65; 19, 22, 27
Fractions	62; 1 63; 13, 17, 30 64; 4, 19, 32 65; 5, 34
Functions	
Linear	65; 6, 12, 15, 28, 29, 32, 38
Notation	63; 30 64; 31
Polynomial	61; 5, 22 62; 9 63; 21 65; 19, 40
Quadratic	62; 26 64; 16, 24 65; 27, 35, 40
Graphs	61; 3, 19, 20 63; 27, 33, 37 64; 2, 18, 24 65; 13
Inequalities	61; 20 62; 29 63; 5, 19 64; 27, 39 65; 10, 13, 23, 31
Logarithms	61; 6, 19, 30 62; 17, 28, 33 63; 5, 30 64; 1, 21 65; 6, 21, 31
Numbers	
Complex	62; 21 64; 34 65; 11
Real	62; 34 64; 7 65; 11, 31
Reciprocal	63; 3 65; 7
Operations	61; 4, 9, 13 62; 2, 27 64; 31 65; 11

Polynomial expansion	61; 5, 7 62; 12 63; 9 64; 20 65; 14
Progression	61; 12, 24, 26 62; 3, 14, 20, 31, 32, 40 63: 16 64; 6, 22, 28, 34 65; 5, 20, 24, 36
Proportion (see also variation)	61; 2, 15, 37 63; 25, 38, 39 64; 13, 26, 29, 35 65; 2, 8, 18, 28, 38
Ratio	see Proportion
Recursive relation	62; 32 64; 31
Roots (of an equation)	61; 29, 33 62; 11, 21, 34 63; 14, 24, 28 64; 7, 8, 30 65; 1, 6, 7, 22
Sequence	see Progression
Series	see Progression
Variation	62; 13 64; 5 65; 37, 38

Geometry

Coordinate geometry	61; 3, 19, 20, 39 62; 19 63; 1, 7, 12, 33 64; 2, 15 65; 9, 13
Plane geometry Affine properties	62; 15 63; 7, 27, 33 65; 4
Angles	61; 8, 25, 27 62; 7, 20 63; 6, 22, 34 64; 36 65; 25, 30
Area	61; 16, 21, 32 62; 6, 16, 18, 37, 39 63; 15 64; 15, 22 65; 16
Circles	61; 11, 38 62; 5, 25 63; 15, 18 64; 13 65; 2, 30, 39
Classification (of figures)	64; 17 65; 25
Congruence	65; 30, 35
Polygons 3 sides	61; 38 62; 7, 23 64; 39 65; 8, 16, 30, 37
4 sides	61; 39 62; 16, 25 63; 32 64; 10, 22, 33 65; 12, 25, 35
n sides ($n > 4$)	61; 27 62; 18, 31 65; 2
Proportion	61; 23, 31 63; 18, 22 65; 8, 12, 36, 37

Pythagorean formula	61; 10, 14, 21, 36 62; 6 63; 10, 25, 35 64; 33, 35, 38 65; 35, 36, 39
Ratio	see Proportion
Similarity	see Proportion

Miscellaneous Topics

Logic	62; 30 63; 26 64; 12 65; 17
Sets	61; 4, 20, 34 62; 28, 29, 33 63; 24 64; 16, 25 65; 29

$$n(i+o) = n\pi$$

$$n \cdot o = 2\pi$$

$$ni = n\pi - 2\pi$$

$$\boxed{i = \frac{(n-2)\pi}{n}}$$

n=3 $60° = \frac{1}{3}\pi$ →4   ✓

n=4 $90° = \frac{\pi}{2}$ →8   ✓

n=5 $108° = \frac{3}{5}\pi$ →20   ✓

n=6 $120° = \frac{2}{3}\pi$ →   ✗

n=7 $128.5° \approx \frac{5}{7}\pi = \frac{900°}{7}$

n=8 $135° = \frac{3}{4}\pi$

$\frac{6}{15}$ $\frac{9}{10}$